History of Science

A Beginner's Guide

ONEWORLD BEGINNER S GUIDES combine an original, inventive, and engaging approach with expert analysis on subjects ranging from art and history to religion and politics, and everything in between. Innovative and affordable, books in the series are perfect for anyone curious about the way the world works and the big ideas of our time.

anarchism
artificial intelligence
the beat generation
biodiversity
bioterror & biowarfare
the brain
the buddha
censorship
christianity
civil liberties
classical music
cloning
cold war
crimes against humanity
criminal psychology
critical thinking
daoism
democracy
dyslexia
energy
engineering
evolution
evolutionary psychology
existentialism
fair trade
feminism
forensic science
french revolution
history of science
humanism
islamic philosophy
journalism
lacan
life in the universe
machiavelli
mafia & organized crime
marx
medieval philosophy
middle east
NATO
oil
the palestine–israeli conflict
philosophy of mind
philosophy of religion
philosophy of science
postmodernism
psychology
quantum physics
the qur'an
racism
the small arms trade
sufism

History of Science

A Beginner's Guide

Sean F. Johnston

ONEWORLD
OXFORD

A Oneworld Paperback Original

Published by Oneworld Publications 2009

A CIP record for this title is available
from the British Library

ISBN 978–1–85168–681–0

Typeset in Jayvee, Trivandrum, India
Cover design by A. Meaden
Printed and bound in Great Britain by
Bell & Bain, Glasgow

Oneworld Publications
185 Banbury Road
Oxford OX2 7AR
England
www.oneworld-publications.com

Contents

Preface

Historical studies of science and technology have acquired immense importance since the Second World War, and especially over the past few decades. Science itself has grown exponentially to involve more activity within living memory than through its previous history. Academics, government policy-makers, businesses, public interest groups and scientists themselves increasingly recognize the crucial role these subjects have had in extending knowledge, driving economies, influencing opinion and shaping culture. Science in the modern world is all-encompassing and contentious. Acronyms jostle for attention and comprehension in newspaper stories: GMOs, BSE, vCJD, WMD. We consume science in the films we watch, the electronic products we buy and the medications we choose. Science subtly determines our perceptions and powers, our lifestyles and longings.

Aspects of what we recognize as science have been part of human cultures since prehistory, though. As these activities have impinged increasingly on the consciousness of scholars and the public, the history of science has attracted further attention and been applied to conflicting purposes. For the British philosopher William Whewell, writing during the early nineteenth century, science (and the people defined by his new term, *scientists*) had long provided the key factor behind intellectual advancement. His *History of the Inductive Sciences*, published in 1837, helped to launch a new discipline. From the Victorian era until the Great Depression, the history of science was marshalled to show the inevitability of material progress. And for both American and Soviet philosophers after the Second World War, the trajectory of science in their nations represented the superiority of their respective political systems. But according to the counterculture ideals of the late 1960s, the history of science revealed a long-standing linkage between scientific knowledge and military and corporate power. Historians today recognize science as a human activity responsible in large part for the culture that we have inherited.

This book introduces both the history of science and the nature of the evolving discipline, and explains why they matter in contemporary society. It seeks to provide a brief historical survey of science and its relationship to culture and, at the same time, to inculcate in its readers an awareness of the changing definition of science and sensitivity to its historical interpretation. This traditionally has been a minefield for educators, with science students sometimes presuming that a history should reveal the progress or even moral improvement of humankind, and humanities students more often presuming tarnished or questionable ideals of what in the twentieth century became a pervasive cultural activity. The tone of the book seeks to give a nuanced appraisal of the evolution of science and rational approaches to the natural world from prehistory to the present day, and its integration into the practices and goals of Western societies over the past three centuries in particular. At the same time – inevitably selective and limited in coverage – it explores the nature of academic knowledge by providing an overview of the evolving discipline of history of science. The 'received wisdom' about past science has changed dramatically in the last generation. This short book seeks to chart those changing directions.

I thank Marsha Filion of Oneworld Publications, who invited me to write this book, my students, who helped shape it, and my wife Libby and sons Daniel and Sam, for their ideas and support.

Sean Johnston

Illustrations

1
Introduction

History of science – past and present

What is history of science? You have picked up this book with expectations, and maybe even unconscious assumptions. Today, more than ever, your assumptions may be different from those of others around you.

More than other forms of history, the history of science has often been written with a purpose, but those purposes and the conclusions they cite are today increasingly questioned. Are you looking forward to reading about geniuses and their life stories? About scientific breakthroughs and inevitable material progress through invention? About the challenging experiments, toil in the face of personal, institutional or military adversity, and the ultimate triumph of the intellect? Or (I hope) something more?

History of science has been all these things, but today strives to be much more. Written by scientists, history can seem self-serving; by philosophers, it can suggest a logical trajectory that is far too clear in retrospect. This potential for misrepresentation can have undesired side-effects: it can encourage unsustainable faith in science's achievements, or provoke unreasonable criticisms of the cases that do not meet the mark, and may deter even bright students from confidently considering science as an attainable career.

But 'misrepresentation' suggests that there is an accurate, objective, official version of the history of science to be told. Surely a succession of precise and indisputable facts will reveal how and why science developed in the way it did? Careful detailing of events is unquestionably important, and historians of science have ever more carefully explored the circumstances surrounding episodes of discovery and invention. But describing large-scale events and their causes is contentious, particularly in broad surveys like this one. Which facts are significant? Which historical personages matter? From the early nineteenth century, for example, Isaac Newton (1642–1727) came to represent in Britain an icon of exceptional and unrivalled genius. In his shadow were others arguably worthy of attention, too: his contemporary Robert Hooke (1635–1703), who advanced microscopy and other experimental sciences, and Newton's rival Gottfried Leibniz (1646–1716), deviser of a more powerful version of calculus. Later scholarship revealed that Newton, as eventual head of the Royal Society, had played an important role in vaunting his own status in England. And in the 1930s – some two centuries after his death – historians began to turn their attention to his vast studies of alchemy and biblical scholarship, neither of which are categorized as science today, but to which Newton devoted equally meticulous attention and probably more of his time. The result was a more nuanced portrait of a complex man.

Nevertheless, the pursuit of great thinkers has been a common thread in history of science. They can serve as models to emulate or to be nurtured. Accounts of exceptional individuals also encourage us to see intellectual development as sparks of inspiration, or – to use a term first popularized during the First World War – as *breakthroughs* that are asserted to be the inevitable result of concerted brain-power. We cannot blame a lack of historical facts for these often misguided popular visions. Albert Einstein (1879–1955) is today an icon of scientific genius. But his latter decades of relatively unproductive science, his role as supporter of left-of-centre causes and his love affairs are less well known, and yet significant to his life's work. And a more recent example still, the American physicist Richard Feynman (1918–1988), has been cast in popular histories of science as a

Figure 1 Richard Feynman, 1975: how did he represent science? (Photo: S. Johnston)

quirky genius, imbued with a unique creativity and wholly unlike his contemporaries (Figure 1). Do such depictions capture the essence of their lives? Do such extraordinary individuals typify, or contradict, the development of science?

Originally seeking to document intellectual advance, history of science has for over two hundred years often been closely associated with philosophy and questions of how knowledge gets refined. The French philosopher Auguste Comte (1798–1857) understood science as an intellectual and historical process. He argued – in ways that would raise the hackles of many scholarly communities today – that mathematical sciences represented the culmination of intellectual progress, carrying humankind from what he and many European contemporaries saw as primitive superstitions and animism to monotheistic theology (both seen as fictitious proto-theories) to metaphysical ('abstract') and then to what he called 'positive knowledge' itself.

His vision of inexorable advance via hard-nosed scientific methods won converts into the twentieth century, and provides the skeletal rationale for some practising scientists today.

With the rise of history of science as an increasingly recognized profession in the twentieth century, some of the assumptions about genius and progress were questioned. Scientists, philosophers and historians increasingly diverged in their views. When examined in detail, intellectual change seemed often to depend on factors that had been neglected. American philosopher of science Thomas Kuhn (1922–1996), in his seminal text *The Structure of Scientific Revolutions* (1962), argued that battling for scientific theories involved not just important facts, but also the perceptions of scientific communities supporting them. Others, like British broadcaster and historian James Burke (1936–) in his ruminations in the television series *Connections* (1978), went so far as to suggest that scientific advance and technological change were quixotic flukes, an unpredictable and unenlightening series of fortuitous juxtapositions of people, places and insights.

Bracketed by Comte and Burke 150 years apart, history of science might seem irrelevant: why study its history, if scientific advance was almost preordained, on the one hand, or completely meaningless and unforecastable, on the other? One reason is that historians and others remained fascinated by the complex episodes, their profound human consequences and attempts to explain their trajectory. A more relevant motivation was that philosophers, sociologists and historians from the 1970s began to focus on the broader factors involved in creating new knowledge at every scale, ranging from the organization of laboratories to public perceptions to national politics. What had captured attention nearly two hundred years earlier as a straightforward and inspirational illustration of humankind's progressive drive has become a rich territory for wide-ranging disciplines.

What is science?

As more reasons for studying the history of science were identified, the foundations themselves came under increasing examination. Historians have been apt to adopt broad, inclusive definitions of their subject. Even modern descriptions cover a lot of ground. As defined by the *Oxford English Dictionary* (Second Edition, 1989), for example, science includes 'knowledge acquired by study', or a 'recognized department of learning' concerned with 'demonstrated truths' or 'observed facts, systematically classified', along with 'trustworthy methods for the discovery of new truths'. And, in a narrower sense, the OED defines its modern usage as 'branches of study that relate to the phenomena of the material universe and their laws'. Not all of these components are necessarily essential for our purposes (dictionaries tend to be heavily weighted towards contemporary usage). But within these dry and seemingly straightforward phrases are hidden dimensions that will be at the heart of this book. What sort of study, for example, has been employed and, indeed, what kind of knowledge is produced? How have truths been demonstrated, and by whom, for what audiences? How are facts best observed and classified, and how have trustworthy methods been developed? And are the answers to these questions obvious to all, or contentious?

Shaking off the firm convictions of the Victorians, the term *science* began to appear increasingly uncertain as a term during the twentieth century, and a timeless definition now seems inadequate. During one relatively brief period – over the sixteenth to eighteenth centuries – profound changes in scientific knowledge and practice occurred, resulting in historians such as Herbert Butterfield (1900–1979) popularizing the term *the scientific revolution*. For others, such as physicist and philosopher Pierre Duhem (1861–1916), the modern form of science began in the late twelfth century, with earlier activities described as *pre-scientific*.

More recent scholarship, e.g. by historian Steven Shapin (1943–), has questioned the amount of discontinuity during the 'revolution'. Some aspects of careful observation and rational explanation can be traced back much further. As I will try to show, the scope and content of science have changed century by century, not just in terms of what we know but also how and what we choose to study, and what we include within it.

The shifting boundaries are important in this. Defining what science *is* can also be aided by seeking a consensus about what science *is not*. As historians have demonstrated – but few science textbooks attest – the borders of science have been repeatedly challenged and adjusted. On one side lies science and, on the other, 'pseudo-science': a field that fails to live up to contemporary norms. Examples abound, and can be important to philosophers and sociologists to explain how new knowledge is assessed and validated, and how new sciences come to be.

Take *phrenology*, for example. During the early 1790s Franz Joseph Gall (1758–1828) devised his new system of brain anatomy and categorization in Vienna. He argued that portions of the human brain were responsible for particular intellectual attributes, and that their relative size was reflected by the shape of the skull. A decade later, accompanied by J. G. Spurzheim, Gall developed his ideas and undertook a successful lecture tour throughout Europe. By 1815, this would-be science of phrenology was attracting harsh criticisms from elite medical journals but attracted further public attention and spurred middle-class men to take up phrenology as a scientific pursuit. Phrenological societies and subject journals, modelled on existing scientific journals, proliferated from the 1820s. The Phrenological Association first met in 1838, mimicking the British Association for the Advancement of Science (from which the phrenologists had been excluded). This professional interest was reflected in popular culture. By mid century, novelists such as Mark Twain (in *Huckleberry Finn*) and Gustave Flaubert (in *Madame Bovary*) were referring to phrenological ideas. Despite such popular interest, phrenology failed to become an established science, and gradually found itself simplified and sidelined by the end of the century as a contentious technique to identify born criminals and to classify human races, and later to associations with sideshow mind-readers. Threads of these ideas nevertheless influenced turn-of-the-century anthropology and twentieth-century neuroscience.

Was this an unjustly persecuted science? Many phrenologists were convinced of it. They cited a clear set of scientific claims (including that 'the mind is composed of distinct, innate faculties' and that 'the shape of the brain is determined by the development of the various organs', and hence 'as the skull takes its shape from the brain, the surface of the skull can be read as an accurate index of psychological aptitudes and tendencies'). By contrast, their first critics argued that phrenologists were not trained medical men and had no recognized qualifications; and, perhaps most damningly, they ridiculed the phrenologists' claim that the mind was entirely contained within the brain, an idea that smacked of materialism (i.e. that natural processes could fully explain living and animate things) – a criticism later levelled at Charles Darwin and his theory of evolution. In an early application of the history of science, the phrenologists complained that they were in the position of Galileo some two centuries earlier, victimized by an established authority that would not recognize the true nature of things!

Level playing fields

Cases like phrenology are fascinating in their own right, but also raise questions for historians of science. To modern eyes, some of the criticisms about phrenologists made by their contemporaries seem misguided. The downfall of phrenology did not depend merely on scientific tests of their claims, but on the ferment of nearly forgotten social factors. And the case was an opportunity to challenge, and shore up, imprecise orthodoxies as much as to attack rival claims.

Such border skirmishes can also reveal complexities of scientific assessment unnoticed, or unmentioned, by practising scientists. The case of the would-be science of *spiritualism*, between the 1850s and 1920s, is a good example of the history of science providing insights to philosophy, sociology and to the practice of science itself. The eighteenth-century Swedish man of science, Emanuel Swedenborg (1688–1772), first conceived the scientific study of spiritualism, which included powers of clairvoyance and communication with spirits. Although supported by certain American Christian sects from the 1840s, the subject flourished from 1848, after John and Margaret Fox and their daughters, Catherine and Margaretta, moved into a house which the girls claimed was haunted. The girls devised a system of communication (shortly after the invention of the Morse code, curiously enough) based on rapping on the walls.

The direct evidence of spirit communication made a sensation. In 1853, the first Spiritualist Church was founded – an example of the continuingly close association between religious and scientific claims. Within two years spiritualism claimed to have two million followers. Like phrenology, the expanding subject had an established set of claims. On the face of it, the claims appear less easily tested than those of phrenology. For example, spiritualism claimed the existence of genuine *mediums*, privileged individuals sensitive to the vibrations of the spirit world (in fact, use of the term 'vibration' itself suggests the links they drew with modern science, which was then exploring wave phenomena in acoustics and optics). Spiritualism claimed that the spirit world was inhabited by spirits retaining the existence and personality of individuals after their death, and who could communicate via mediums. And it supported these claims via the phenomena of the *séance*, a special laboratory-like setting of subdued lighting and multiple observers. Spiritualists claimed that, in the séance, a spirit could manifest itself or materialize animate objects from *ectoplasm*, or could send messages by *mechanical writing*, rapping or vocalization, all via the specially adept medium. Unlike phrenology, however, these claims appealed to scientists, particularly physicists, psychologists and philosophers. A number of British intellectuals founded the Society for Psychical Research in 1882 to explore the phenomena.

By 1900, numerous mediums and séances had been studied, and some 11,000 pages of reports were produced. Some investigators, like scientist William Crookes (1832–1919, and known particularly for the Crookes tube, or early cathode-ray apparatus), became convinced of the genuine psychic abilities of certain mediums. A growing number of scientists, however, came to distrust the claims owing to the difficulty in reproducing them, and because of some exposed frauds. By the early 1920s, spiritualism was declining strongly in popularity, possibly because of the public distrust in its reliability following the many attempts by families to communicate with the recent dead of the First World War.

Unlike phrenology, spiritualism as a claimed science did not die. It retained a coterie of followers, although most were no longer scientists. In a sense, it was reborn as a new would-be science: in 1927, Joseph Banks Rhine (1895–1980) and his wife, Dr Louisa E. Rhine (1891–1983), came to the psychology department of North Carolina's Duke University to study psychic phenomena, which they recast as the science of *parapsychology*. Gone were ectoplasm and materialization, replaced with *extra-sensory perception* (ESP) and *psychokinesis*, part of a larger category of phenomena the researchers dubbed *psi* phenomena. Séances were replaced by laboratory experiments, standard apparatus such as Zener cards and, later, statistical analysis and computers. In the intervening decades, parapsychology has attracted scientific criticism based on its elusive results and subtleties of interpretation. It survives in a number of university departments, although increasingly relegated to phenomena of *mis*perception rather than in the original sense of *extrasensory* perception. The evolution of parapsychology mirrors that of psychology itself during the twentieth century: becoming more mathematical, reliant on instrumentation and refined in its experimental protocols – and so capable, in principle, of detecting fainter effects. Is this a science-in-the-making, or another case justifiably to be sidelined? The history of science provides useful comparisons and contrasts.

What is a scientist?

Just as the definition of *science* challenges our preconceptions so, too, does the word *scientist*. What does it conjure up in your mind? Probably you imagine a male, quite possibly in a white lab coat, and perhaps working in a government or corporate laboratory. This vision is a recent one, scarcely a half-century old. Looking further back, the environment changes: from sponsored research to smaller-scale, more individualistic studies and – in some locales – a gentlemanly pursuit. As we travel back in our imaginary time machine to the early nineteenth century, the scientist abruptly disappears altogether, because the term was coined only in 1833 by British philosopher William Whewell (1794–1866). The novel word encapsulated a new vision of what these experts were, and it was not universally applauded. Michael Faraday (1791–1867), for example, detested the idea of commercial gain as a motive for seeking scientific knowledge. Earlier *men of science* or *natural philosophers* shared a different collection of intellectual, professional and religious attributes than their modern counterparts. So, the history of science is populated with a changing set of actors through the centuries, and the activity is longer-lived than any stable set of practitioners.

Just as the individuals mutated in form, so too has their public perception. A growing number of scientists were criticized for their views or effects on religion, such as Isaac Newton in the seventeenth century and Charles Darwin in the nineteenth. Portrayals during the twentieth century vacillated from characterizing them as eccentric but creative eggheads to admirable problem solvers, to disturbingly unreliable and powerful figures in society. For much of that time, scientific practitioners have been both praised and criticized for their relationship with society – another important trait documented by historians of science. But these attributions also challenge us: was Fritz Haber (1868–1934), the Nobel-Prize winning chemist who invented the commercially important industrial process for synthesizing ammonia but also responsible for German gas warfare during the First World War, a model scientist? According to Germany of the 1910s, unquestionably so; twenty years later, however, Haber, a Jew, was forced to emigrate from Hitler's Germany, where another poison that he had invented as a potent insecticide was later used to gas some of his own relatives. By exploring the context of such events, history of science can reveal dimensions that have more recently been explored, such as the connections between science and ethics, politics and national identity.

Where is this book headed?

This short book has three goals. First, as suggested by the examples above, it aims to survey the mutating meaning and influence of science in its changing cultural contexts. It will focus on the ideas, practices, innovations, events, individuals, groups and institutions that shaped science to trace the trajectory of this intellectual and social activity over time. What is science, and how has it worked? What have its products been? And how did this special form of knowledge come to be wielded as a potent tool in modern Western societies? It will impart the flavour of the ideas, events and effects that, in the past generation, have become widely recognized as significant. Reflecting the expanding diversity and quantity of scientific activity, attention will be weighted toward the more recent past.

The second goal is to justify this approach, and to explain why this particular analysis is both relevant to current culture and fair-minded (warning: seeking 'truth' – both historical and scientific – is an ambition that has been increasingly challenged by philosophers). As befits a *Beginner's Guide*, this book will sample not just historical case studies, but also many of the scholarly themes that have shaped the history of science. This collection of changing understandings – some shared, but others mutually exclusive – should give a sense of this growing discipline.

The third goal is closely related to the first two. It seeks to answer the question, why should we care about the history of science? I have hinted above at some of its important uses, at least for a few academic disciplines. But I will argue that – much more than recognized in the recent past, and in far more varied ways – the 'back story' or subtext of history of science is relevant to multiple audiences and previously excluded communities. These include the general public concerned about, or enamoured with, contemporary scientific and technological change, and specific populations previously portrayed as peripheral to science, such as colonial societies, entire cultures (e.g. that of China) and women as a gender. As C. P. Snow (1905–1980) observed in the mid-twentieth century, the 'two cultures' of Science and the Humanities should be brought together for mutual benefit. The history of science builds such a bridge. It crosses disciplines and makes links as significant for media and gender studies as for students of the classics or engineering. The study of the history of knowledge, craft skills and innovative tools has long been a foundation of Liberal Arts and many Humanities programmes in universities, but is revitalized by new academic perspectives. And for all readers, history of science provides the power not merely to better understand important aspects of our past, but to inform our present-day judgments, too.

2

Big ideas and compelling approaches

The broad historical development of science can be tracked through a bewildering variety of social and cultural contexts, and a surge of intellectual ideas, craft practices, measuring instruments and expressions of human curiosity. At the heart of these entwined aspects, though, lies the question of how some ancient cultures came to adopt and value activities that we would now class as part of science. This chapter, like those to follow, will combine a roughly chronological coverage with relevant themes. It focuses on aspects that were periodically identified as useful by different cultures: regularities of the natural world, practical utility and the adoption of rationalism. This perspective will sketch the confidence repeatedly placed in such systematic features of knowledge, and how they were shaped and applied within and between distinct cultures.

Early sources of knowledge: studying the sky

Where to begin? I will cast the net as widely as possible to suggest that some of what we understand as science has been a feature of many human cultures, although frequently in forms that blur and challenge modern distinctions. We cannot confirm this without a thorough survey, but can at least argue that these activities are nothing wholly new or unique to a single people. When, then, did cultures begin to trust reliable truths about their natural world based on systematic observation and other organized methods?

Archaeological evidence is limited for tracing the cultural activities of ancient peoples, but there are indications that celestial phenomena provided a certain commonality between prehistoric and subsequent cultures. In modern urban environments the sky is largely forgotten and unobservable, a source of complex phenomena that we no longer notice. By contrast, all pre-industrial cultures had a rich collection of sky phenomena to study, explain and use – a set of activities that can be mapped onto science to a degree.

Phenomenon: an observation or occurrence that is unexplained or unusual. A phenomenon for *you* may not be a phenomenon for your neighbour, if she has a satisfying explanation for the occurrence.

As the most obvious and useful night-time phenomenon, the moon's peregrinations would have been obvious to any people with access to clear skies. For Paleolithic peoples reliant on good weather, seasonal animal migrations and wild crops, the phases of the moon, repeating in a cycle of about twenty-nine or thirty days, provided a particularly obvious time-marker: the period between quarter-phases is just over seven days, our modern week.

Besides its evident phases it had other unique properties. It moved through the background of stars, some twenty-six of its diameters every 24 hours. On occasion, it underwent a sudden darkening (the lunar eclipse). And, perhaps most remarkably of all, it appeared to be precisely the same size as the sun. It is

Figure 2 Moon phases as a celestial clock (l-r): New Moon, First Quarter, Full Moon and Last Quarter (Photo: S. Johnston)

worth interrogating your own knowledge and explanations of these characteristics. For example, how long does a lunar eclipse last, and can it occur at any phase of the moon? Why is the moon such a perfect size-match for the sun, and why do the two travel at such different rates through the background of stars?

An array of such celestial phenomena were observed and later recorded, some regular and obvious, and others subtle and complex. In distinction to this familiar but mysterious celestial body, all ancient peoples could readily see that the sun had its own peculiar properties. On rare occasions it disappeared dramatically (the solar eclipse), but in a way very unlike the moon's disappearance. Less obviously, except for peoples far from the equator, the sun describes a seasonal trajectory through the sky. Rising roughly in the east and setting towards the west, its course in the sky was observed to be highest in summer and lowest in winter. The length of the day varied with the seasons. And the point on the horizon at which the sun rose and set varied regularly, ranging furthest north in summer and swinging southward in winter.

While the awareness and use of such phenomena is only hinted for Paleolithic communities (possibly because sparse surviving materials such as notched bones tell us little about craft practices), there is ample evidence that Neolithic peoples closely monitored the sun's characteristics. Large-scale stone constructions became widespread across Northern Europe and the Mediterranean during the Neolithic period. Stonehenge (*c*3000 BC), the most famous of these, can be understood at least in part as a structure that encapsulates this annual periodicity. By the alignment of its massive stones, it marks the extremes of the sun's trajectory to show the position of sunrise at midsummer (and sunset at midwinter), and possibly other celestial phenomena. It is not unique: sighting channels through buildings in Neolithic settlements as far apart as Newgrange in Ireland (*c*.3300 BC) and certain pyramids in Egypt (*c*.2500 BC) appear to have served similar purposes. (The field of *archaeoastronomy* has since the 1960s sought to assess such evidence.) It is significant that such creations married craft skills (stone working and land engineering) with technical design to devise techniques of visual observation.

These prehistoric works also hint at the social significance of astronomical observations. Given the exorbitant societal cost of building such structures, they undoubtedly held cultural relevance beyond mere calendar marking. Agricultural benefits likely played a role: in harsh northern climates where a successful growing season could be determined by the week of planting and harvest, a precise knowledge of the year could be crucial for a marginal agrarian community. These skilfully designed structures serve to illustrate the long human awareness of regularities of the physical environment, and growing interest and ability to systematically observe and record them.

Further evidence for intellectual curiosity linked to cultural interpretation lies in the subtlety of phenomena that were studied. We know from the later written records of several ancient civilizations that consistently long-term interest and documentation became widespread. Besides the sun and the moon, the stars themselves were perceived to have complex attributes. They formed recognizable and unchanging patterns (*constellations*), and passed overhead at a rate subtly different from the sun's motion, shifting slightly from night to night by about two diameters of the sun. All but a handful of them moved as a unit, in rigid lock-step. The five exceptions – some brighter than all but the moon and the sun, and others extremely dim – had individualistic properties. Each of the special stars moved through the sky unlike the others, inhabiting a band through the background of stars, and somehow associated with the sun. And unlike every other object in the sky, including the sun and moon, three of special stars (our Mars, Jupiter and Saturn) would occasionally reverse their apparent motion, seeming to travel backwards against the stellar background (later dubbed *retrograde motion*). But there were other matchless phenomena in the sky: the faint band of light crossing the sky seldom visible from cities today but known to us as the Milky Way; and fainter smudges and arcs almost never seen by modern eyes. In short, the night sky was a treasure trove of arcane and diverse phenomena, each rigidly reliable in an apparently endless series of details and part of the quotidian experience of every adult and child of ancient cultures.

Explaining phenomena

All these attributes repeatedly encouraged not merely systematic and extended observation, but explanation, too. In what is present-day Iraq, the Sumerians (*c.*3500 BC) recorded celestial observations, and associated the celestial bodies that moved against the stellar background – first the sun and moon, and later the five special stars – with deities. This astral theology became the basis of religion for their successors, the Babylonians, and also a motivation for further systematic observation. Observatories became associated with temples, and valued the detection of unusual celestial phenomena for the purposes of divination. One such surviving document is the Venus tablet of

Ammisaduqa, a copy of Bronze Age observations of the planet Venus dating to *c.*1600 BC. Babylonian observations revealed long-term regularities between planetary and lunar motions, some many decades long, and permitted predictions of some eclipses. While deeply embedded in theological meaning, such precise forecasting suggests the pragmatic benefits that accrued from the careful observation and analysis, and also demonstrates the societal investment that was made to permit such long-term activities.

Nevertheless, written records reveal less about the co-development and spread of craft knowledge. Metal working – the mining, smelting and casting of gold, silver, copper, bronze and later iron – was an art deeply reliant on careful observation, experimentation and transmission to new practitioners. These skills and their associated technologies and applications represent a relatively undocumented thread of science in ancient societies.

Given the early association between written records and astronomical observations in particular, it appears that celestial phenomena were an important common factor in many early societies. This source of phenomena was unlike earthly occurrences: it provided regular events that could be seen, remembered and documented over a lifetime, and it displayed a spectrum of unique properties graduated in their ubiquity and complexity and suitable for every degree of examination. For many cultures, an almost palpable *system* of the heavens suggested a division between the earthly and the celestial, and yet a close connection with earthly concerns such as agriculture. It is not unreasonable to claim that this source of information was a crucial and widely recurring cultural motivation for the development of observational methods, rational thought, metaphysical speculation and explanation for ancient peoples, just as we can more confidently assert that it was for later cultures. In brief, at least some activities and methods today understood as science – although used and allied with other cultural practices in unfamiliar ways – appear to have been a common feature of even the earliest human societies.

Rationalism: the Greek heritage

By preserving oral, written and craft traditions, succeeding generations and cultures were able to usefully accumulate knowledge of a powerful kind. One key skill was careful observation, necessary to perceive subtle or long-term regularities. But while systematic observation and recording may have been a common feature of many societies, other features that have been characterized as 'scientific' were not. One of those approaches was a particular form of explanation based on reasoning and further observation. This *rationalism* became a feature particularly promoted by influential thinkers in early Greece. Socrates (470–399 BC), his student Plato (427–347 BC) and Plato's student Aristotle (384–322 BC) established an approach to reasoning about the natural and human-made world that much later became the basis of Western philosophy.

> **Rationalism**: the view that reliable knowledge, and ultimate truth, is obtained through reason, evidence and logic.

Plato, in his text *The Republic*, wrote, 'astronomy compels the soul to look upwards and leads us from this world to another'. Aristotle's writings illustrate the motivation and satisfaction of explaining phenomena. The phenomenology of the sky – i.e. its collection of unexplained observations – ranged from obvious to subtle.

Greek reasoning about such phenomena sought to relate the evidence provided by our senses to persuasive rational causes. In most cases, this reasoning about manifest experience was allied closely to 'common sense'. This alliance between direct experience and intuitively reasonable ideas to explain cause and effect is compelling. Consider your own understandings of the natural world. Now that personal food gathering, navigation and time keeping have been superseded, astronomy no longer has practical importance for most of us. Most astronomical phenomena remain just that to us: we cannot give an adequate explanation for why they happen, although we trust that astronomers can. This trust in authority is reasonable, and suggests why Greek explanations were convincing to successive generations and cultures. However, each individual also continues to construct explanations about events based on his or her own experience.

Consider a 'thought experiment' to explore your own mental models of reality. (This is intended to explore your sources of trust and to appreciate the achievements of earlier thinkers, not to teach an official version of truth.) Imagine that a cannon ball is shot straight up into the air from a moving train (Figure 3). Where does it land, at point A, or B or C along the track (the points are fixed points on the ground)?

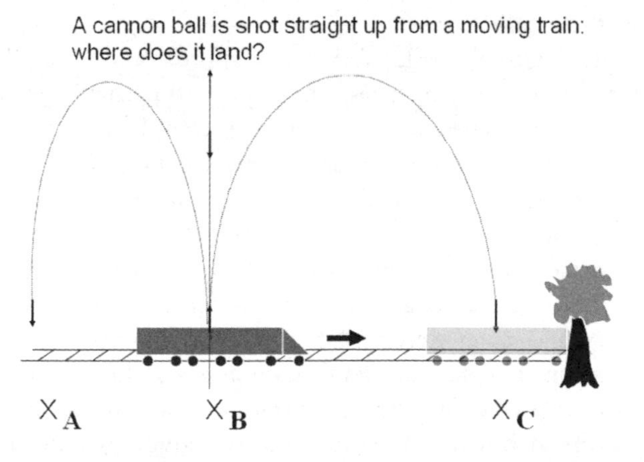

Figure 3 Thought experiment: projectile on a train (S. Johnston)

Answers from casual surveys vary, with B or C being the most common. Aristotle argued for B: once the ball is shot into the air, he would reason, it is free to follow its natural tendency, which is to fall straight down to earth from where it had been fired. The accepted answer today, though, is C: the cannon ball, shot straight up, will travel with the moving train to land on it again. In case this doesn't accord with your common sense, consider a variation of it based more directly on your experience: imagine that you are in the train carriage, throwing an orange up and down in your hand. Does the orange hang in space as the train moves under it, to hit the rear wall of the carriage? No: it moves along, as you do, with the train and lands neatly in your hand again.

The thought experiment reveals many of us to be Aristotelians at heart. Such examples suggest that what we believe, and the way we conceive the world, often depend on something other than merely looking for ourselves. Aristotle's explanations became authoritative, though, for reasons beyond seeming to confirm common sense. They satisfyingly categorized the natural world and its attributes.

Both Plato and Aristotle sought to describe universal truths about the natural world. Aristotle's approach was to identify *essences*, or general properties, that explained particular examples of phenomena. These essences could, in turn, be used to explain new observations. For example, he proposed that four elements were sufficient to describe the material world. Earth, air, fire and water (in idealized pure forms) combined in different proportions to yield every earthly substance. These elements each had particular qualities, or attributes. Both earth and fire were dry, while air and water were wet; fire and air were hot, while earth and water were cold. Each had a natural place of rest as well, and an associated tendency. Fire's natural place was above the earth, and so fire-rich materials would tend to move upwards to reach that place. Smoke, for example, being mainly a mixture of fire and air, would rise into the sky. Bubbles in water would migrate up to the surface. On the other hand, rocks, being made up primarily of the element 'earth', would fall to the ground, their own natural place of rest. Just as twentieth-century physicists were attracted by the elegant simplicity of atomic theory based on a few fundamental particles, so too were followers of Aristotle deeply satisfied by his handful of elements and essences.

Such Greek principles covered not just the inanimate world, but living things as well. For instance Theophrastus (371–287 BC), a contemporary of Aristotle, had written influential treatises on botany and compiled classifications of varieties of animals, plants and minerals. Aristotle, too, had categorized hundreds of species, interpreting their forms and behaviours according to fundamental principles. Just as his four elements of earth, air, fire and water explained the nature of the inanimate world, his doctrine of 'four basic qualities' (hot, cold, wet and dry) explained concepts of health and illness in terms of opposite and complementary forces, and based corresponding treatments on the need for balance and stability. More generally, and according to similar principles, Greek and Roman physicians had followed the writings of Hippocrates (*c.*460–*c.*370 BC) to seek a balance of four *humors* (black bile, yellow bile, phlegm and blood). According to Aristotle, these affected personality and diagnosis (a 'sanguine' or 'phlegmatic' patient being attributed to an excess of the respective humor) as well as treatment (e.g. blood-letting).

Aristotle made few connections between mathematics and the inanimate and animate natural world. One exception was his understanding of falling objects. He claimed that the quantity of motion would be proportional to the object's weight (implying that heavy objects fall faster) and inversely proportional to the density of the medium through which it fell (explaining why a rock falling through air moves faster than the same rock sinking in a pond). An incidental consequence of this idea is that empty space cannot exist: if it did, objects would move infinitely fast through it, and so would instantly fill up any void that appeared.

What about other forms of motion, though? According to Aristotle, a thrown rock has two forms of motion. First, the throw imposes an unnatural motion on the rock. Second, when this imposed action has been 'used up', the rock will then pursue its natural motion, which is to fall straight down to earth.

A fifth element, *aether* (known as *quintessence* during the mediaeval period) was the constituent of all heavenly bodies. This literally unearthly substance, according to Aristotle, filled the region of space above the earth. It was pure and unchangeable, a property confirmed by the ageless constancy of heavenly bodies. Aether had no qualities of heat or humidity, and its natural tendency was to move in a circle – a quality consistent with another obvious property of heavenly bodies.

The reduction of physical and biological phenomena to a handful of elegant concepts was seductive and pleasing to Greek scholars. It was not a static and universally accepted understanding, however. Plato, Aristotle's teacher, had developed similar concepts but founded on his own reasoning that emphasized geometry as a guiding principle. The four earthly elements, for example, corresponded to four aesthetically pleasing regular solids (the tetrahedron, cube, octahedron and icosahedron, having four, six, eight and twenty faces, respectively). The 'pointiest' of them (the tetrahedron) was associated with the stabbing heat of fire; the cube, being the least spherical, was associated with the element earth. The fifth, and celestial, element corresponded to the fifth Platonic solid, the dodecahedron. Astronomical bodies travelled in circles, the most perfect geometrical form.

In the centuries following Hippocrates, Socrates, Plato and Aristotle, Greek understandings of the natural world continued to evolve, but largely within the satisfying framework defined by their predecessors. Aristotle had argued that the aether made celestial bodies incorruptible and unchangeable. Based on the records of earlier cultures available to the Greeks, astronomical knowledge did indeed appear stable and worthy of its investment of intellectual labour. The most persuasive and useful refinement of Aristotelian knowledge about the heavens was the *Almagest* written by Claudius Ptolemy (83–161), living in Roman Egypt, around AD 150. Originally a Greek work entitled the *Mathematical Treatise* or *Great Treatise*, the *Almagest* is a corruption of its Arabic title, al-kitabu-l-mijisti or 'Great Book'. He was known for other important works, too: his *Geographia* compiled all known geographical and cartographical information of the Roman empire, and his *Tetrabiblos*, or 'four books', became the most popular text on astrology of his time. Ptolemy also investigated the mathematics of music and the properties of light.

For the *Almagest*, Ptolemy used contemporary observations alongside ancient Greek and Babylonian astronomical records to devise an elaborate form of Aristotle's heavens. The sophistication is suggested by its complexity and by its ability to model celestial observations extremely well, while retaining Aristotle's fundamental principles. Ptolemy's system relied on an elegant series of circular motions. The motions of all the heavenly bodies, including the irregular motions of the five known planets, could be

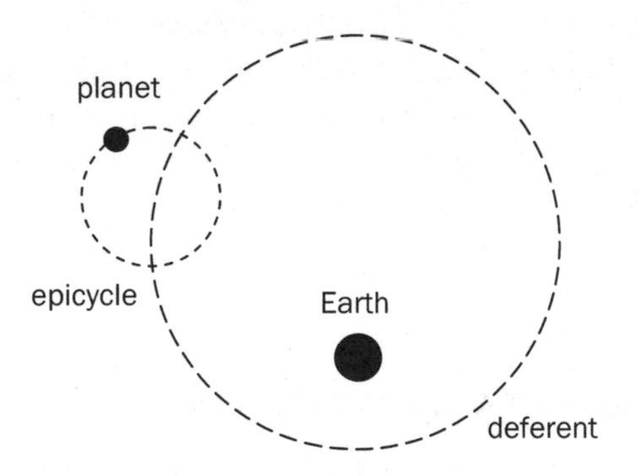

Figure 4 Ptolemaic system of planetary motion (S. Johnston)

explained by their movements along circles known as *epicycles* that moved along larger circles known as *deferents* (Figure 4).

Theory and theology

The fact that Greek thought is remembered suggests the power of scientific explanation to transcend cultures. The *Almagest* was copied and disseminated widely over the next fifteen hundred years. Its calculation methods were refined and corrected to accord with further astronomical observations. Its longevity is remarkable, particularly when we consider the immense cultural changes that occurred between AD 150 and the seventeenth century. Whole societies were influenced, merged and superseded. New religious systems – notably Christianity and Islam – swept the Middle East and beyond, incorporating elements of Greek concepts into their theology. The mutation of scientific practice through cultural change has become an important focus for historians of science.

The Romans added relatively little to Greek natural philosophy, but Greek thought was insinuated into Roman scholarship, principally in the form of compilations and encyclopedic texts. A prominent example is the *Naturalis Historia* (*Natural History*) of Pliny the Elder (23–79), a seminal source for mediaeval scholars and one of the largest surviving Roman works. After the Roman world was officially Christianized in the fourth century the distinct world views of Greek philosophy, Christian theology and Roman law were increasingly reconciled.

Christian accommodation

New institutions provided a conduit for scientific knowledge, too. Monasteries became the storehouses of ancient learning in the post-Roman world. Although solitary monks had grown to be a popular religious tradition from the third century in Egypt, collections of monks in monasteries became significant social institutions throughout Europe a few centuries later. Monastic communities were centres of industry, agriculture and intellectual preservation. In what is now north-eastern England, for example, the Venerable Bede (672–735) taught at a Benedictine monastery and wrote the first account of its inhabitants, *The Ecclesiastical History of the English People* (731).

As the principal surviving compendia of knowledge, the Roman encyclopedias were an important source for monastic scholarship. Nevertheless, the texts copied and disseminated to monastic libraries by scribes such as Bede provided an incomplete sampling of Greek knowledge. The Hellenistic portion of the Roman Empire had lost its political connections with the West when the empire was divided into eastern and western halves under separate rulers. The advance of Christianity introduced further divisions, with Eastern Christianity adopting distinct theological foundations. As a result, knowledge of Greek – the language of ancient scholarship – declined in the West but was retained in the Byzantine Empire. Through Europe, even Latin translations of Greek texts were sparse, and were limited mainly to the encyclopedias that summarized Greek handbooks. Much of Plato and most of Aristotle had been lost.

Indeed, modern understandings of the scientific knowledge of past cultures is still far from complete, as historians have long realized. This is even more true for practical sources and implementations of the knowledge, which could not be copied by scribes and stored safely. One illustration is the 'Antikythera mechanism' discovered in 1900 in a Greek shipwreck. Dated to the first century BC, the corroded bronze device is a sophisticated calculating mechanism. Inscribed with a lengthy set of instructions for use as an astronomical instrument, it could be employed to predict the positions of the sun, moon and planets and rising and setting of known stars. The complex geared machine is an entirely unique survivor; no written Greek texts suggest anything equivalent, and its sophistication as a scientific instrument has been compared to clocks and orreries of the eighteenth century. Because of this void in the historical record, historians of Greek science unfortunately have been forced to focus on the culture's intellectual development rather than on the related and co-evolving craft knowledge and artisanal skills.

Even so, some Greek science was preserved and eventually transmitted to subsequent societies in ways that other cultural heritage was not. Ancient Indian scholarship, which had developed mathematics, medicine and astronomy centuries before the time of Christ, was only indirectly accessible in the West after the division of the Roman Empire. Other cultures were too distant, or too isolated, from the Mediterranean world to influence its scientific development significantly. The cultural products of China, so remote for travellers until the Renaissance, could be transmitted only via a long chain of traders. This form of 'Chinese whispers' limited knowledge-transfer: any fact was embellished, often corrupted or even wholly invented on its long voyage; travellers' tales, more often than not, provided little of value for science. Complex craft skills were as likely to be lost as to be successfully adapted. Even material products such as porcelain (exported from China to the Middle East by around AD 600) often arrived devoid of the techniques that had fashioned them. And completely unknown until the sixteenth century were the cultures of Central America, which developed their own culturally distinctive astronomy, mathematics and architectural mechanics. Even then, profiteering adventurers and Christian missionaries overlooked, and sometimes actively destroyed, their records and constructions. The transmission of knowledge between cultures, then, was a hit-or-miss affair.

The transition from the ancient world to the European

culture of the early Middle Ages also carried a significant judgment concerning rational knowledge. Knowledge itself became Christianized and, in certain respects, devalued. The ultimate sources of knowledge were rechannelled from ancient authorities to the Bible. In the process, the accepted methods of discovery and evaluation (so-called *epistemology*) were reshaped.

Epistemology: The philosophical study of knowledge and systems of knowledge and reliable methods of discovering it.

Early Christianity, like the Roman Empire in which it became established, placed rather little emphasis on scientific knowledge. Practically useful information was more valued. Ptolemy's works, for instance, were studied because of their importance in determining the annually changing date of Easter, a source of acrimony between the early Christian churches of different regions.

Beyond such utilitarian theological concerns, Christianity also added spiritual, sacramental and symbolic truths to Greek rationality. Aristotelian science could be used to illustrate the natural order suggested by Biblical interpretation. Earth and God's people were the centre of the universe, and overseen from Heaven, the domain of God. This addition of an extra sphere to Aristotle's model was the only significant change to his cosmology (Figure 5).

These views of the natural world extended to nature, biology and purpose, too. As historian Lynn White Jr (1907–1987) has argued, the Christian view emphasized humans as separate creations of God distinct from the 'background' environment. Human domination of nature was interpreted as God's will, and the act of creation and human life had the purpose of fulfilling God's plan. This, some ecologists have argued, formed the basis of Western attitudes towards the natural environment and

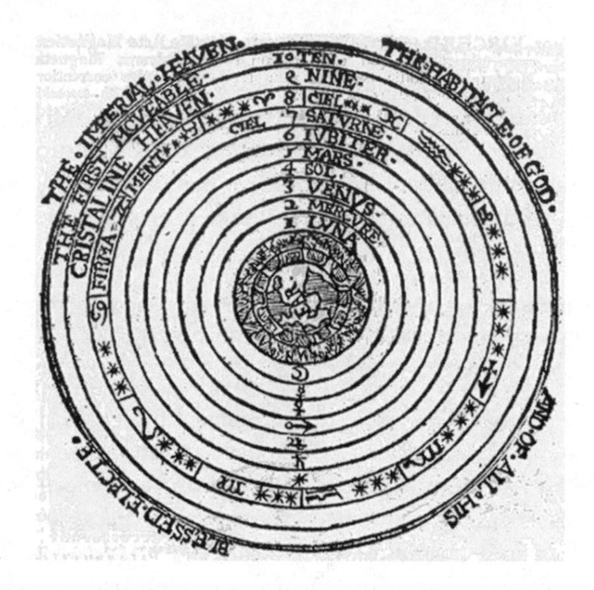

Figure 5 The geocentric universe as illustrated in *The Geomancie of Maister Christopher Cattan*, 1591

sustainability. According to this Christian interpretation, then, the natural world was a carefully constructed stage on which religiously meaningful episodes were played out. Based on such a world view, careful investigation of phenomena, and deep pondering of cause and effect, seemed to serve little purpose other than illustrating the glory of the creator.

Arabic science

Christians were not the only religious communities to adopt and adapt Greek natural philosophy, nor even consistently the most important. Natural philosophers of the Middle East, in fact, transferred, adapted and used Greek knowledge more effectively than early European Christians, especially when empowered by the rise of Islam. Its culture was a conduit for ancient knowledge and a source of scientific innovation. By contrasting such different cultural expressions of science, historians can better understand how context affected its development.

The death of Muhammad in AD 632 was followed by a period of disputes within the Muslim community concerning religious and political leadership. Within a century, though, Islam expanded rapidly and extended through the Middle East, North Africa and Spain. And between the eighth and thirteenth centuries, the so-called Golden Age of Islam consolidated this religious culture.

The cultural environment fostered by Islam produced a distinctive form of scientific knowledge. One reason for dissimilar Islamic and Christian directions was the different sources on which they drew. Where early Christian scholars inherited the concise books compiled by Roman encyclopedists, Muslim conquests brought both ancient and fresh knowledge from Persia, India and Greece.

In addition, Arabic science mutated and extended these ancient authorities. Unlike contemporary interpretations of the Christian Bible, the Qur'an emphasized the usefulness of experience and empirical observation. But where the Greeks had highlighted rationality, scholars in the Islamic world stressed experiment. The growth of experimental techniques was earlier and more rapid in the Middle East than in Europe. The field of investigation was broad and productive, ranging from astronomical observatories (based on naked-eye observations, of course) to surgical techniques, and to chemical manipulations that permitted the study and use of new medicines and new materials. The development of mathematics applied to experimental observation was another clear distinction from Greek methods.

An important example of this new approach was the work of Ibn al-Haytham (965–1039, known as Alhazen). His *Book of*

Optics (1021) outlines a methodical approach very much like modern scientific methods. Unguided observation would be followed by statement of the problem, formulation of an hypothesis and its testing by experiment. The results would be analyzed to draw a conclusion, and the work would be published. Alhazen's *Optics* used this approach to explore and convincingly explain optical phenomena in a way that earlier natural philosophers had not.

During its Golden Age, Arabic science yielded practical and conceptual results beyond European achievements. As this knowledge spread gradually to Europe its role in fostering Western science was profound. In the borderlands between mediaeval Christianity and Islam (notably Islamic Spain and Sicily), European scholars gained access to many dozens of Arabic texts and translated them into Latin. The result was a flood of new knowledge during the twelfth century from the ancient cultures of Greece, India and Persia along with the new astronomy, alchemy and medicine of the Arab world. This was seldom a neutral transmission of knowledge, however: as intermediaries, Arab scholars edited, interpreted and extended their sources. Terms coined by Islamic mathematics (e.g. algebra, algorithm), chemistry (alcohol) and astronomy (Aldebaran, Altair, Betelgeuse) are examples of this heritage. Less obvious but even more important was the introduction of the Arabic number system (itself an import to the Arab world from Indian mathematicians), which made computation less laborious than the Roman numerals used throughout Europe until the Middle Ages.

As the new-found Arabic knowledge seeped into Europe, scholars began to closely examine lost texts and to re-evaluate their content. An important example of the re-packaging and extension of Greek and Arabic ideas is the work of Johannes de Sacrobosco, or John of Holywood (1195–1256). A Canon in an Augustinian abbey in the British Isles (most likely in what is now southern Scotland), Sacrobosco was educated at Oxford and later became professor of mathematics at one of the first mediaeval universities, at Paris. He taught mathematical subjects, and his gravestone labelled him as a 'computist', or time-reckoner. Sacrobosco became one of the most popular authors of the Middle Ages with his *de Sphaera* ('*On the sphere*', *c.*1230). The book, an elementary treatise on astronomy, explained Ptolemy's system and incorporated ideas from Arabic astronomy. For over two hundred and fifty years it was a popular hand-copied university text. From 1472 it was reproduced by printing presses, reaching over ninety editions by the late seventeenth century.

From the twelfth century, then, Christian, Islamic and more ancient cultural sources fertilized active re-examination of scientific knowledge. The merger meant more rapid change, but instability and conflict, too.

Magic and method

The focus so far has been on the evolution of some recognizable attributes of modern science in particular cultures, especially confidence in rationalism, empiricism and experimental technique. But this new basis for trust developed in an environment of competing ideas. Historians of science have increasingly paid attention to these alternate forms of knowledge and shown their close links with science. This broadening of attention has helped to define what science 'is not', as discussed in chapter 1, as well as the roots of scientific practices in other forms of knowledge.

Although the Greeks had developed a rational philosophy – a way of understanding the world by geometry, mathematics and mechanical explanations – and Islamic research and application emphasized experimental exploration, craft skills and artisanal knowledge, there was still a deep trust and confidence by many of their scholars, and within other cultures, in other kinds of explanation. These other forms of belief were more or less mystical, sometimes religious, and sometimes based on hidden varieties of knowledge. There was a gradual shift of allegiance from what could broadly be called 'magic' to 'science'. Two of the most important and popular beliefs that grew alongside early science were the subjects of astrology and alchemy. Their evolution illustrates the wider cultural context, and how the definitions and activities of science gradually were shaped.

Scientific practices developed piecemeal. As suggested by the importance of Greek culture and its early Christian and Islamic adaptations and extensions, the progressive extension of knowledge was not guaranteed. The subjects of interest, their explanations and their social application were distinct to each culture. And even within the mediaeval Christian societies of Europe, the framing of knowledge was variously configured by rival intellectual communities.

One important distinction lay in the nature of explanation itself. This branching drew upon Greek roots: followers of so-called *Neo-Platonism* distinguished themselves from *Aristotelian* scholars. Neo-Platonism gained supporters in the fourth-century Greek world (although this term for them was coined only in the eighteenth century). Neo-Platonists revived interest in Plato's ideas and sought to merge them with other philosophical and theological traditions. Their ideas influenced some early Christian and Islamic thought, notably the writings of Augustine of Hippo (354–430). Among their essential claims were that the universe is ordered hierarchically under a single god, or 'the One'; the abstract forms making up the universe were illustrations of that divine creator; and this framework of creation was eternal and unchanging. Augustine was influential in defining Western Christianity and, with it, notions of the natural world.

When Christian and Islamic traditions intersected from the twelfth century, Aristotle's philosophy (largely favoured by Islamic scholars) began to exert a stronger influence than Plato's thought in Europe. Christian scholars increasingly explored empirical methods, making observations and drawing conclusions from them. One of the notable examples was Roger Bacon (1214–1294). A Franciscan friar, Bacon read the work of Alhazen and embraced his experimental approach. Undoubtedly inspired by Islamic writers, Bacon wrote on a wide range of scientific questions including optics, calendar reform and alchemy.

Astrology and alchemy

The interplay of Platonic and Aristotelian ideas can be traced by their influence on distinct bodies of knowledge. The roots of astrology, as noted earlier, trace from prehistory. The body of knowledge, and the texts that disseminated it, became increasingly elaborate and formalized during the late Middle Ages. Two traditions competed for ascendancy. So-called 'natural astrology' could be understood as an extension of Aristotelian science. Its concern was with mapping the correlation between celestial events and earthly natural events – for example weather, agricultural yield and large-scale human occurrences such as epidemics or even the fate of nations. By contrast, 'judicial astrology' dealt with celestial predictions for important individuals. Both employed astronomical observation, calculation and analysis (or interpretation). *Nativities* were celestial maps at the time of an individual's birth; *horaries* were similar maps at the time a prediction was made. Both were a form of 'horoscope', or time-picture at a significant moment. Analysis relied on more than the mere positions of celestial bodies, though. Each astronomical object carried a number of distinct attributes which could determine the interpretation. For example, the planets were associated with masculine and feminine aspects; with wet and dry, hot and cold; with certain occupations and professions; with certain colours, illnesses and bodily organs. There was an elaborate set of connections between causes and effects. So the position of a planet in a certain constellation, or 'house', could be interpreted according to these associations. The criticisms of this seemingly rationally minded enterprise concerned the difficulties of reliable analysis based on this complexity. Interpretation, it appeared, relied on a carefully trained astrologer sensitive to multiple layers of meaning. Unpeeling this celestial onion made it qualitatively different than, say, performing Ptolemaic calculations of planetary motions. Astrological observation was linked to explanation and underlying theory in a manner that was difficult to discern for non-practitioners. The entire subject was fenced by intellectual, mystical and social barriers.

Alchemy was more directly an import from the Middle East. The name suggests Arabic origins in practical skills (*al-khem* meaning black earth) or Greek roots (*chemeia*, the art of making metal ingots, or *chumeia*, extracting juices or infusions from plants). Early alchemy – i.e. as a collection of techniques for separating and combining materials – was certainly practised in Egypt, Greece and India much earlier than Islamic culture but, in any case, this art developed significantly under Islamic scholars. The specialized equipment they invented to distil, extract, filter and fuse chemical substances is recognizably familiar to that found in modern chemistry laboratories. The earliest expression of alchemy, then, as well as its enduring legacy as a part of chemistry, was as a subject based firmly in the practical arts.

Alchemy reached Europe as part of the wave of imported Arabic texts during the twelfth century. Their alchemical writings were descriptions of chemical procedures and apparatus developed over the previous millennium in the Middle East. The texts ended up in the hands of the few who could read, write and study: men of religious orders. These were translated but also extended. Robert de Chester's *De compositione alchemiae* (1144) is an early example. During the thirteenth century, treatises on alchemy were written by the Bishops of Lincoln and of Regensburg. Roger Bacon, also a cleric, was an example of this new wave of Christian scholars inspired by Islamic natural philosophy. With further extensions, alchemy became more closely associated with ambitious aspirations and with magic. In the following century, though, the church's tolerance waned. Franciscan and Dominican orders were instructed not to teach the subject, and during the fourteenth century alchemy was classed as a heresy. Lumped with magicians and wizards in handbooks for inquisitors, alchemists were deemed dangerous because their mystical abilities seemed to transcend or bypass God as the keeper of mysteries. This irreligious form of knowledge was perceived as both suspicious and powerful, and practitioners of alchemy protected it with highly encoded texts. The knowledge was restricted to adepts – those who had studied the subject and absorbed its jargon and symbolism. The dense layers of meaning are hinted by an extract from the *Pretiosissimum Donum Dei* (*Precious Gift of God*, 1475):

> I am black of white, and red of white, and citrine of red, and certainly I am a true sayer and not liar. And know ye that this red of the art is the Crow which in the blackness of the night and in the clearness of the day flyeth without wings. Of the bitterness in his throat the color is taken, from his body the redness, and from his back pure water. Understand the gift of God, receive it and hide it from all unwise philosophers, for it is not hidden from the caverns of the metals, which stone is mineral, and animal, shining colors, or high hill, and an open sea.

Expertise in absorbing and applying these writings could be gained by various routes, none of which coincided with mediaeval universities or monasteries. An alchemist could be introduced to the subject by personal instruction from a master through apprenticeship; or, he might gain personal revelation by living a particularly ascetic or contemplative life; or, he might gradually acquire knowledge through private experimentation or assiduous study of the texts. Just as astrology imbued celestial objects with many properties, the alchemical terms, descriptions and illustrations carried layers of obscure symbolism.

Occult: from Latin *occultus*, meaning hidden. Knowledge distinct from conventional sources such as theology and science.

Neo-Platonic ideas were revitalized in Europe during the Renaissance. The circulation of translated Arabic sources and a new interest in antiquity stimulated scholars of the fifteenth century. Over the next two hundred years, some bodies of knowledge – especially astrology and alchemy – became more elaborate, more formalized and layered with additions of symbolism. Each became an established and recognized craft – a complex craft for which meaning was malleable and subtle.

This account suggests that distinct communities became allied with certain forms of knowledge. An important example of such groups in seventeenth-century England were the Parliamentarians ('Roundheads'), who opposed the Royalist supporters of King Charles ('Cavaliers'). William Lilly (1602–1681), the best known astrologer of the day, popularized the subject by writing the first English language text on the subject, *Christian Astrology* (1647). As a movement, the Puritan roundheads placed confidence in Lilly's astrological predictions of planned battles against the Cavaliers. Even though the King also consulted astrologers, Lilly and his craft became identified with a 'democratizing' tendency. Not only did he forecast the defeat of the King by less hierarchical opponents, but Lilly also provided these common people with knowledge of an intellectually dangerous and seditious kind. As a result, English support for astrology developed clear religious and political overtones. This sectarian support was largely responsible for the eventual decline of astrology in England. When Charles II, successor to the deposed Charles I, regained the throne after the Restoration in 1660, his supporters linked astrology with disrespect for the authority of the monarch. Royalists labelled astrology a species of profane divination founded on mere imaginary suppositions. The King's sponsoring of the new Royal Society that year hints at his snubbing of astrologers and the social rise of science.

Alchemy did not decline as astrology did. Instead, it was transformed during the seventeenth century from a mystical craft to a more publicly disseminated science. Two members of the Royal Society (discussed further in chapter 3) were active alchemists: Robert Boyle (1627–1691), one of its founders, and Isaac Newton, later its President. Both were transitional figures who conformed to traditional alchemical methods while also exploring new mathematical extensions. In fact, historians seeking to rebalance the accounts have referred to Newton as 'the last of the magicians' and 'last alchemist'. In his book *The Sceptical Chymist* (1661), Boyle rejected the leading scientific theories of his day and started the list of elements which are still recognized today. He also formulated a mathematical law relating to the volume and pressure of gases.

The two subjects of alchemy and chemistry had distinct foundations, concepts and goals. Alchemy incorporated aspects of Neo-Platonism and came to incorporate subtle mystical interpretations. An important guiding concept of mediaeval alchemy was the notion of a substance that would convert one form of material into another. This 'oil of philosophers' or 'powder of projection' or 'philosopher's stone' accumulated a range of properties as the subject developed. It could transmute 'base metals' – tin, lead, iron – into 'noble metals' – silver and gold. It might also enhance health or even extend life or potentially provide immortality.

By contrast, the new chemistry was promoted as being more empirical. Less mired in symbolism and obscure interpretation, it distanced itself from theological and metaphysical links. Relatively rapidly and publicly, chemistry revised concepts and achieved useful results. But this sketchy comparison is biased to some extent; an alchemist of the seventeenth century would surely disagree with my assessment. He would probably argue that the goals of chemistry are narrow. Alchemy sought to explain the interconnectedness of all aspects of the natural world, linking physical materials, forms of life, medicines and religious truths with philosophical meaning and purpose. The 'occult' and 'obscurantist' philosophy described in the writings of alchemists, he might claim, reflected this deep holism, while chemists limit themselves perversely to the crudest phenomena visible to unskilled observers. And, while a philosopher's stone was never publicly demonstrated to wide audiences, the achievements and goals of alchemy were greater than those of chemistry. Our alchemist would claim that his subject tied together the world and made it sensible as a coherent system. A less partisan critic might even examine late alchemy and early chemistry side-by-side for a fairer comparison. Not all claims of the new chemistry were self-evident. One of the early investigations of the Royal Society involved air pumps. One interpretation was that it created a new, artificially created entity – a 'vacuum' – although its existence and plausibility were actively disputed. The philosophical debates centred on the interpretations of ingenious and varied experiments performed by investigators who became adapt at the specialized experimental apparatus. And during the following century, chemists identified a series of gases (nitrogen, oxygen, chlorine) for which the properties were just as subtle to interpret as the materials isolated by alchemists.

For contemporary observers, these disputes between old and new forms of knowledge were difficult to resolve. For historians of science, they challenge our ability to think 'symmetrically' and fairly about competing knowledge claims. The confrontation between 'magic' and 'science' – through intellectual, theological, social and political skirmishes – was played out over more than two centuries. In retrospect, it has been called revolutionary.

3

One damned revolution after another?

Historical changes of pace, or radical transitions of thought and practice, may be more evident to subsequent generations than while they are occurring. Such periods may even be identified by the way later observers themselves explain and partition the world. During the twentieth century several such historical periods were identified and studied extensively, and historians have disputed the degree to which scientific change is continuous or abrupt. This chapter samples several episodes, each distinct in its own way: periods subsequently described as the 'scientific revolution', the 'Industrial Revolution', the Enlightenment and the Darwinian revolution. And another case, the geology of Alfred Wegener, will introduce the problems of identifying intellectual revolutions. In their various ways, these complex social and intellectual transitions promoted new concepts, new practices and new ways of viewing the world.

The scientific revolution

The term *scientific revolution* has been sprinkled through the previous chapters of this text. Nevertheless, this is a historical change that is clearest in retrospect. Its strength and discontinuity were disputed by scholars during the twentieth century. The weight of recent evidence, though, is that the gradual development of mediaeval practices and understandings of the natural world altered in profound ways. During the sixteenth and seventeenth centuries and into the eighteenth, a cluster of profound hypotheses, new methods of discovery and new styles of spreading knowledge combined to yield a distinctly different world view for Europeans. Over some two centuries what came to be called the 'new science' or 'new philosophy' gained adherents until it had replaced one orthodoxy with another.

This revolution had wider historical precedents. It can be seen as an outcome of the so-called *Renaissance* ('rebirth'), beginning in fourteenth-century Italy. This cultural movement revived interest in classical sources and led, for example, to patronage for versatile artists and inventors such as Leonardo da Vinci (1452–1519) (not to be confused with Galileo Galilei [1564–1642], discussed at greater length below). Spreading through northern Europe, this network of ideas and practices fostered new confidence in scholarship and revived interest in actively investigating the natural world.

Alongside this new emphasis on the *studia humanis* – the study of humanity – was attention to practical invention: together, gunpowder, movable-type printing and the compass (all developed and used in China centuries earlier) provided social power, dissemination of knowledge and new confidence in exploration and trade. Exploration, particularly to the Americas from the early sixteenth century, established new trade routes and revitalized European economies. It also revealed the 'new world' of the Americas, and along with it new cultures, new species and new natural phenomena, convincing scholars that classical authorities were neither adequate nor reliable.

The new approaches that became associated with this period deserve individual examination so that we can assess their origins and significance. And, supported by the topics of chapter 2, it is important to trace those subjects that declined in importance. As historian John Henry has argued, one of the 'revolutionary' aspects of the scientific revolution was the way in which subjects were redefined. Bodies of knowledge were picked apart and reassembled into new configurations. Disciplinary boundaries shifted to become more *or less* inclusive. This is a key point: the evolution of scientific ideas during this time cannot be seen as merely the accumulation of discoveries, but instead as the reassessment of all that was known. This re-evaluation altered more than merely understandings of how the natural world functioned. The new philosophy also transformed ideas about how knowledge should be acquired and disseminated; how knowledge should be used; and how humankind related to God. We will pick up the theological dimensions in chapter 4, and will focus on the others here.

Consolidating knowledge

New ideas become easier to introduce when there already are alternatives available. The last chapter emphasized the role of Greek writings in mediaeval thought. Through the accident of the preservation of his works and selective filtering to suit the needs of the early Christian church, Aristotle was understood as the authority regarding natural philosophy. Renaissance rediscoveries of other Greek philosophical writings challenged this autonomy, though. The most important among them, at least for natural philosophy, were the works of Plato (*c.*427–347 BC), which stressed the importance of mathematics in explaining nature. Geometry, it was claimed, offered more than just craft methods and calculating skills, but mystical layers of natural understandings. Revived during the Renaissance, these ideas stressed the importance of reason for understanding all aspects of the natural world. This emphasis on rationality challenged, to some extent, prevailing interpretations of Aristotelianism, which stressed the importance of sensory experience in verifying knowledge.

> **Empiricism**: the view that reliable knowledge and ultimate truth are obtained through sensory experience alone.

At the same time, texts on magic were also rehabilitated. Like mathematics, magic had evolved continuously from the ancient world but had been largely excluded from mediaeval natural philosophy. Together these contrasting and ancient forms of knowledge encouraged a challenging of old authorities along with fresh evaluations. Natural philosophy was enriched and destabilized by rediscovered texts. New explorations sought a merger of old ideas with new knowledge, and were likely to apply a combination of rationalism and empiricism to questions about the natural world.

Established knowledge was not immune to this re-examination. This book began with prehistoric astronomy to suggest that human attributes – attention to regularities and the desire to explain them – were combined with environmental features (clear skies), social needs (hunting and, later, agriculture) and artisanal skills (e.g. stone working for alignment markers) to yield some of the features we commonly recognize as science. Astronomy has been widely identified as a key activity for the changes during the scientific revolution, too.

The most visible challenge to ancient authority was the questioning of Claudius Ptolemy's astronomy. Ptolemy (around AD 83–161) had devised a mathematical model of the heavens that explained celestial observations very successfully. It allowed him to explain not just the movements of the stars, sun and moon, but also the wandering of the five known planets against the starry background, and even their variations in brightness. His model had been used, refined and adjusted as necessary by generations of astronomers over a period of thirteen centuries – surely an excellent test of its worth. For Ptolemy, undoubtedly, the system represented a vision of celestial reality.

Realism: the philosophical view that explanations can be refined to accurately describe the true nature of physical reality.

Even so, Ptolemy's system had built-in contradictions which opposed the teachings of Aristotle. Where Aristotle had understood all heavenly bodies to travel in perfect circles around the earth, Ptolemy required that some of them move in more complex ways. Both he and Aristotle agreed that circular motion was the only natural movement for objects in the heavens, but Ptolemy's planets moved in circles (epicycles) around invisible centres, and those centres moved along larger circles (deferents) (see Figure 4). Without an explanation for the cause of these 'unnatural' motions, Ptolemy's system had to be accepted as an accurate mathematical model rather than a self-evident Aristotelian demonstration.

Reconstructing world views

With the passage of many centuries of observation, it is not surprising that deficiencies in Ptolemy's system cropped up. These could be repaired by adjusting the values he had defined for the size of epicycles and deferents. More precise observations required further adjustments, and astronomers progressively added further epicycles to better approximate their observations. Although this tweaking deviated from his originally simple and elegant scheme, it was seen as an elaboration, not as a condemnation, of his concepts. Ptolemy's model could flex and adapt, intellectually speaking, because mediaeval scholars gradually came to interpret it as a tool rather than as a physical reality.

> **Instrumentalism**: the philosophical approach of treating any accepted fact or theory as a working hypothesis or provisional truth, i.e. as merely an *instrument* or tool in order to discover further knowledge.

Seen as an instrument to make mathematical predictions, the Ptolemaic model became disconnected from Aristotle's vision of the heavens. In this way cosmology and predictive astronomy were divorced; one revealed the nature of things while the other tracked celestial movements.

The procedure of fitting models to observational data also encouraged relaxation about considering any deeper significance. In this intellectual environment, the Polish astronomer Nicolaus Copernicus (1473–1543) sought ways of improving the accuracy of the calculations of celestial observations. His mathematical solution, though, was to conceive a new astronomical system quite unlike Ptolemy's. Where Ptolemy imagined the earth motionless at the centre of his complex model, Copernicus required that the earth move in a circle around a motionless sun. Ptolemy's model was *geocentric* while Copernicus's was *heliocentric*. Viewed instrumentally, the Copernican model was unproblematic. It achieved similar accuracy with fewer circles, and so was significantly easier to calculate in practice: a convenient pragmatic tool. Viewed from a realist perspective, though, it was disturbingly provocative. Aristotle and Ptolemy – and academic orthodoxy – argued for a motionless earth. A moving earth was expected to produce effects which were not seen, and so the Copernican model could not possibly represent reality. Even so, as he showed in the preface to his book *De Revolutionibus* (*On the Revolutions of the*

Heavenly Spheres, 1543), Copernicus eventually came to believe that its mathematical elegance must faithfully represent physical truth. The calculating tool and cosmological models had to be reconciled.

But this mathematical edict posed serious problems for natural philosophy. Recall, for example, the thought experiment of chapter 2, where we imagined an object shot upwards from a moving train. What would happen if an object were launched upward from a rotating earth? According to Aristotle, the object would fall straight down *while the earth rotated under it* To all of us watching from the surface of the moving planet, the rising and then falling object would seem to slant sideways, falling far off to the west as it was left behind. More obviously, we would feel strong winds as the earth rotated, and flying birds would be left behind as soon as they lifted off the ground. Aristotle's physics demonstrated that the earth must be motionless, and so the Copernican model was at best a convenient fiction.

The Copernican view of the heavens also challenged more fundamental aspects of Aristotelianism. Without the earth at the centre, how could the natural motions of the 'sublunary' sphere (the region between earth and moon) be reconciled with the very different natural motions of the 'superlunary' sphere? Circular motion was natural for the 'fifth essence' (quintessence) above the moon, but vertical motion prevailed below it; how could up and down now be defined?

So, the reception of *De Revolutionibus* was mixed. The idea of the sun at the centre of the universe was rejected out of hand by virtually everyone, theologian and natural philosopher alike. The book was widely admired for its mathematical elegance and sophistication, though. The result was rather a schizophrenic approach to belief. For most, the Copernican system was understood as a mathematical convenience that did not represent reality. It was merely a calculating tool – an arithmetic short-cut – while Aristotle's cosmology depicted the universe as it truly was. Where Copernicus and his followers gained confidence in realism, his opponents more conservatively accepted the model only tentatively as a working fiction or means to an end. Just as for Ptolemy's model based on epicycles, the audience divided into realists and instrumentalists.

While most considered that such mathematical models demanded no grounding in reality, others sought compromise. Tycho Brahe (1546–1601), for example, was able to match his own meticulous celestial observations to a model that called for a motionless earth but circled by the sun, around which all other planets orbited. This model enabled the reconciliation of Aristotelian physics on earth with a version of Copernicus's model in the heavens beyond. His assistant, Johannes Kepler (1571–1630) reverted to Copernican ideas, but added significant nuances. He demonstrated that planetary orbits are actually ellipses, not circles, and proposed that their sizes are mysteriously determined by geometrical relationships inspired by the concepts of Plato.

These astronomers also contributed to dismantling Aristotle's cosmology by arguing that each of the planets moved independently, rather than along a defined curve or fixed on a crystalline sphere. On the other hand, in place of Aristotle's satisfying natural circles, they proposed no mechanism for the planetary orbits but provided strong pressure for a linkage between astronomical observations and physical models.

Galileo Galilei (1564–1642), a supporter of Copernican explanations, made his name as a professor of mathematics. Like the astronomers Copernicus, Brahe and Kepler, his career choice was lower in the intellectual hierarchy than natural philosophers of his day and, like them, he sought to unite his field with natural philosophy. Galileo's experiments, observations and theorizing challenged Aristotelian concepts and promoted a union of mathematical explanation and empirical knowledge.

Some of Galileo's work may seem unremarkable to modern eyes, but all of it was provocative to prevailing understandings.

In his studies of motion (kinematics), for example, he determined by experiment that all objects fall with the same acceleration, independent of their weight. This contradicted Aristotle's views that heavier objects fall fastest. Similarly, Galileo showed that any motion can be understood as the combination of separate motions; a cannon ball, for example, moves in a parabola because this is the combination of downward acceleration plus a constant horizontal velocity provided by the cannon. Together, these two experimental findings provide the modern explanations for the thought experiment of chapter 2 and the conundrum of why we are not blown over by strong winds as the earth rotates. Just as importantly, they brought the Copernican explanations down to earth.

As the first person to apply the newly invented telescope to astronomy, Galileo also marshalled damning evidence against Aristotle's claims. Galileo's telescopic observations, plainly described in his *Sidereus Nuncius* (*The Stellar Messenger*, 1610), challenged the Aristotelian account of the heavens. His observations of Venus revealed a changing crescent like the moon's; Copernicus's model explained this by the changing positions of the two planets in relation to the sun and earth. Galileo observed four moons orbiting Jupiter, again a contradiction of Aristotelian and Ptolemaic understandings, but allowed in the new, more liberal, cosmology. And his viewing of the moon revealed it to be not perfect and unchangeable, but unsettlingly earth-like, with rugged mountains and plains. Two decades later, Galileo's book *Dialogue Concerning the Two Chief World Systems* (1632) contrasted the Copernican and Aristotelian world views according to their differing principles of natural philosophy. Expressed as a debate between scholars, it provided a rational discussion of evidence and theory, persuasively arguing that Copernicanism was straightforward, logical and factual.

Galileo's claims were not obvious to all, though. His assertions were dependent on a new and unfamiliar scientific instrument: the telescope. The telescope itself was novel and not adequately validated to convince critics and peers. Viewing through it required different techniques than naked-eye astronomy, and before arguing about its observations, it was necessary to argue its veracity. Why should the telescope work in both the sublunary and superlunary spheres, when those domains were so different? And, in an age when witchcraft was understood as a serious danger, some argued that the telescope was a seductive and misleading device possibly inspired by the devil. As discussed in chapter 4, Galileo's unaccommodating stance also made enemies with certain members of the Italian church, leading eventually to censure of his ideas and limits to his personal freedom.

Disputing mathematics and reality

One of the successes of the new philosophers was the gradual integration of mathematics with natural philosophy. A kind of two-tier system had existed among scholars before them. On the one hand were the philosophers, who theorized about the form of the natural world and the causes of things, a branch of knowledge dubbed *episteme* by the Greeks. On the other were mathematicians and astronomers, who devised calculation schemes and sought the solutions for useful goals. These scholars, often based at universities, found themselves lumped with practical users of arithmetic and geometry such as surveyors, navigators and carpenters. This body of knowledge, known as *techne* for the Greeks, was considered a form of craftsmanship associated with lower intellectual and social status. According to this traditional categorization, numbers and mathematical relationships were pale illustrations of the nature of things; natural philosophers, by contrast, dealt with their essences. For this reason, Galileo's appointment to the chairs of mathematics at Pisa and then Padua were steps on a ladder of success. His achievement – both intellectual and social – was to make his mathematics relevant to natural philosophy by analyzing his experiments mathematically. His threat to Catholic teaching was a consequence of this relevance: Galileo argued that the Copernican system was not merely a convenient description of observations, but also represented reality. The calculating trick had conjured up a new world view.

This heightened status for mathematics became an element in a new culture of scientific practice. What seventeenth-century natural philosophers called the 'mechanical philosophy' or the 'new philosophy' had specific characteristics that set it apart from the old forms of scholarship. Besides being open to the mathematization of nature, the new philosophers relied increasingly on careful direct observation instead of relying on ancient authority. As Galileo had illustrated with careful experiments timing the motion of balls accelerating down ramps and of pendulums swinging, observations were more valuable when simplified. Artificial experiments could isolate a phenomenon to be explored methodically and quantitatively.

Alongside confidence in experimentation, the new philosophers also emphasized the importance of shared information. Members of the new scientific societies such as the Royal Society often met to observe an experiment collectively, discussing what they saw and how it could be explained. The shared experience, they argued, could be less subjective, more perceptive and fertile than chance individual observation (*expérience*, in fact, is still the French term for *experiment*).

Reconceiving life

Collective viewing and shared experiences proved effective and could be multiplied, in effect, by straightforward written descriptions of such events. Descriptive texts and illustrations consequently became an important feature of the new science, and nowhere more importantly than in the life sciences. (For this reason, historians of science have long had a particular interest in such texts).

A key figure was Andreas Vesalius (1514–1564), a physician and anatomist. His influential book, *De Humani Corporis Fabrica* (*On the Workings of the Human Body*, 1543), explained anatomy through detailed and relatively accurate illustrations. And as a professor of anatomy at the University of Padua, he performed public dissections of cadavers for his teaching. These witnessed demonstrations provided an unconventional route to knowledge. They bypassed the received authorities on biology such as the texts of Aristotle and the Roman physician Galen (*c.*129–200), and turned to the craft knowledge of surgery, which had long had a lower status than the ancient book-learning. Rather like Galileo's astronomical observations challenging Aristotle, Vesalius's dissections challenged many details of Galen.

Like the new accounts of the physical sciences, biological explanations became more compelling. Dissection grew in popularity as a central method in the teaching of medicine and for further biological study, and such experimental knowledge encouraged further observation and refinement. Among the less reliable anatomical details inferred by Vesalius were understandings concerning the flow of blood. Vesalius conceived that perforations in the wall of the heart could explain the flow of blood through the body. His Paduan successors proposed different routes of circulation, culminating in the careful experimental observations of William Harvey (1578–1657). In some respects, Harvey learned an Aristotelian approach to knowledge. His experimental studies aimed not only to describe, but to explain causes. Even more than Vesalius, Harvey broke with tradition by studying animal anatomy alongside its human counterpart to gain a broader perspective on life forms. These goals and approaches altered the status of biology (and its texts) from the descriptions of natural history to the realm of natural philosophy. Harvey's synthesis of the evidence confronted the then-current understandings. He argued that the circulation of blood occurred through two loops: the first carried blood to the lungs and the second carried blood to the bodily organs. Galen had argued that two types of blood – venous and arterial – originated in the liver and heart, respectively, and were consumed by the organs of the body. According to Harvey, though, the veins carried blood to the heart, and the direction of flow was ensured by valves within them. Most radically, perhaps, Harvey claimed that the liver played no active role in circulation, and that the heart acted as a pump to push, not suck, blood through the arteries in an endless closed loop. The analysis was summarized in his book *De Motu Cordis et Sanguinis in Animalibus* (*On the Motion of the Heart and Blood in Animals*, 1628), again underpinned by careful illustrations.

Sharing knowledge and assessing truth

The precision and lifelike detail provided in the texts of the anatomists at Padua were a clean break with past traditions. The earlier mediaeval texts on natural philosophy had been richer, in a sense. The most popular descriptions of nature were so-called *bestiaries*, wedded with theological understandings. These were a form of compendium or catalogue of animal species to be found in the natural world, combined with a moral lesson. Many were founded on ancient tales with a Christian gloss, just as Aristotle had provided the backbone for mediaeval Christian philosophy. For example, the tale of the beaver described in Aesop's *Fables* (Greece, sixth century BC) provided a meaningful allegory. According to the bestiaries, the beaver, when hunted and cornered, would castrate itself. Devoid of its testicles (which were valued for making medicine), the beaver was of no economic interest to hunters and so could escape. The lesson to readers was clear: by living a chaste life free of vice, a man may escape the devil. In a similar way, the pelican illustrated the qualities of a doting parent, and the lion, as king of the beasts, was a noble creature that would spare the weak or the captive.

By mixing moral instruction with facts of dubious accuracy, natural philosophy was stretched to fit an unaccommodating world view. For the new philosophers, such accounts were worse than useless. The knowledge provided in bestiaries was unreliable, having been imperfectly reproduced by a chain of scribes since antiquity. In the process, the information had been corrupted or embellished fancifully, and was frequently augmented with travellers' tales. The illustrations accompanying the texts were often symbolic rather than naturalistic, and drawn by copyists who had never viewed the animals themselves.

The new philosophers levelled a different criticism at the books written by other contemporaries, the astrologers and alchemists. As described in chapter 2, practitioners of alchemy sought to explain and restrict their art to adepts. Texts were embedded in mysticism and symbolic description. Alchemical allegory was much deeper and sophisticated than any bestiary, because it was intended to be understood only by experts who held the key to interpretation. One consequence was that comparison of alchemical claims was uncertain and time consuming: descriptions could generate rival interpretations; detail was inadequate, and the texts themselves were intentionally obscure and couched in secrecy. For many of the new philosophers of the seventeenth century, the practices and texts of alchemy were at best inefficient and at worst useless (a notable exception being Isaac Newton, who spent considerably more time in collecting, reading and basing experiments on alchemical texts than on mathematical physics – although he did so in secret and without any published legacy).

Unlike the bestiaries, astrological and alchemical texts, the new scientific journals of the seventeenth century sought to describe observations in plain language unencumbered by symbolism. Descriptions and arguments would be direct and clear, and would require no advanced education to comprehend and decode theological or mystical allusions. Such texts and experiments would be accessible to wider audiences, and would promote the rapid advance of knowledge.

Perhaps the best exemplar of this new culture was Robert Hooke (1635–1703). Studying at Oxford, Hooke became an assistant to Robert Boyle (1627–1691) and constructed and experimented with early air pumps. With the founding of the Royal Society in 1660, Hooke was appointed Curator of Experiments, both of his own design and as recommended by members of the Society. The experiments ranged over the curiosities of the time, including the nature of air, the physiology of breathing and the nature of gravitation. *Micrographia*, his meticulous text containing descriptions and illustrations of microscopic observations, was published to great acclaim in 1665. Its large format and accurate annotated pictures – some of which even folded out – of fleas, plant cells and other microscopic marvels were a revolutionary contrast to the depictions in the old bestiaries. At the centre of Hooke's competences were his manual and inventive skills, often focused on the art of instrument design. As Isaac Newton, too, realized in his invention of the reflecting telescope, artisanal knowledge was crucial to observation and experiment.

The new philosophy, then, promoted a stylistic and philosophical lurch. It incorporated new conceptions of the inanimate and living world, new methods of discovering its order, and new styles of describing it. Rejecting existing styles of description, its texts disseminated a new conscious standard of thinking and writing, and a new discourse about the world and the place of humans within it. The pared-down descriptions in scientific journals were written by practitioners, not scribes or draftsmen. Texts strove to be objective; the writer became merely a careful observer and reporter, and sought to be almost invisible in the process. Between the lines, they argued that we discover principally by observing, and this experience should be public or shared. Written in simple language, accounts could allow repetition of staged experiments. Through the texts, readers became virtual witnesses.

It is important to acknowledge, though, that this new scientific culture intentionally sacrificed attributes that had a long and valued tradition. The mysticism associated with astrology and alchemy were an expression of the interconnectedness of nature. Their practitioners disputed that nature could be picked apart and isolated by artificial experiments. As noted in chapter 2, the broad and entwined understandings of alchemy and astrology were replaced by a narrower focus to pursue more restricted goals. The new philosophers traded an inefficient but satisfying holism for a tailored assault on knowledge.

Interrogating nature: scientific knowledge and power

As suggested in the previous sections, knowledge about the natural world has been distinctly configured by different cultures and each human generation. But the relationship between rational knowledge and humans' capacity to control their environment was first emphasized by Francis Bacon (1561–1626). A contemporary of William Shakespeare and rising to the position of both Attorney General and Lord Chancellor under King James I in England, Bacon promoted a methodical approach to the discovery and application of new knowledge. The limitless horizon for this organized form of science was outlined in his utopian *New Atlantis* (1626). In this work of fiction, Bacon depicted 'Saloman's House', a scientific college on an isolated island in which research was conducted for the good of the wider community. Information gathered from around the world was brought back to be analyzed, expanded upon and applied by teams of specialists.

Bacon was promoting more than a system for intellectual expansion. He related knowledge itself to the improvement of society and national status, and suggested that this was consistent with both Christian ideals and human curiosity. His utopian Salomon's House was dedicated not just to discovery but also to applications of knowledge. Inventions, Bacon argued, could transform human society. The idealism underlying the engineered form of society and its public institutions depicted in *New Atlantis* bears some resemblance to the aims of the first scientific societies a generation later.

As noted earlier, such ideas challenged religious and philosophical convictions. In a world that was understood to be susceptible to miracles, why should men waste their time investigating regularities which God could alter or bypass at whim? The acquisition and application of knowledge about the natural world challenged His authority. Bacon's notions were simultaneously exciting and unsettling to different audiences.

One of the best-known attacks on Bacon's approach, and of the new philosophers who followed it, was a chapter in *Gulliver's Travels* (1726) by Jonathan Swift. Gulliver, a stranded ship's captain, visits Laputa, a rocky island flying in the clouds by magnetic levitation. The island is controlled and inhabited by savants having heads literally in the clouds. They communicate by mathematical formulas and music and serve food in the geometrical shapes of Platonic solids. Nevertheless, the intellectuals are so impractical that they are unable to eat, bathe or acquire practical clothing, and require servants to provide these mundane aspects of life. The satire is even more scathing in discussing the object of their labours. The Laputian scholars study subjects that contradict common sense. Despite being founded on mathematical logic, the projects of their Academy are contrary to reason. Houses are to be built from the roof downwards; the colours of paint are to be mixed by blind men. Their research includes methods of extracting sunbeams from cucumbers, turning ice into gunpowder, and even excrement into food – perverse goals in a topsy-turvy world.

The island's name, incidentally, means *the whore* in Spanish (*la puta*), hinting that Swift may have found such scientific investigation not only ridiculous but also morally suspect. The ruler of the floating island could, in fact, wage war by threatening to shade underlying territories and deny them sun and rain, and even literally to crush a rebelling town. In the hands of its unworldly but powerful scholars, Laputa suggested the threat that the blind scientific method might pose.

Swift modelled Laputa's academics on the Royal Society of his day. The book criticized the cost, practicability and goals of scientific exploration. The futility of seeking practical achievements from such preposterous and elaborate research seemed, to many, a fair evaluation of natural philosophy. Charles II, patron of the Royal Society, was no doubt impressed by its members' attempts to discover a reliable means of measuring longitude at sea, but is reported to have found their discussions of the weight of air both comical and ridiculously trivial. Even at the tail end of the scientific revolution, then, cultural authority proved elusive for natural philosophers.

The Industrial Revolution and craft knowledge

During the early seventeenth century, Francis Bacon had promoted very effectively the notion of science applied to human betterment. He also inspired goal-directed invention as a natural purpose for scientific knowledge.

In his *Novum Organum* (1620) Bacon cited printing, gunpowder and the compass as inventions that had transformed literature, warfare and navigation. And in his earlier work, *Advancement of Learning* (1605), Bacon had forecast the fields in which a methodical scientific approach might provide answers. He suggested, for example, that new powers and inventions could prolong or extend life (a goal of contemporary alchemists), improve agriculture by 'making rich components for the earth' and create 'new Threds and new Stuffs'. Some powers, such as 'Instruments of Destruction as of War and Poison', would suit national ends. Others sound even more familiar to modern ears: the ability to alter 'Complexions, and Fatness and Lean-ness', producing 'Exhilaration of the Spirits and putting them in good order' and providing 'Greater Pleasures of the Senses'. If such pharmacological delights sound appealing, yet other powers remain still beyond reach today, but no less sought: 'Encreasing and Exhaltation of Intellectual Parts', 'The raising of Tempests' and even 'the Making of New Species'.

Active application of knowledge was the key: Bacon noted that 'men ought to know that … it is only for God and angels to be spectators'. Bacon's aphorism 'knowledge is power' (1597) much later became a guiding principle of industry.

Over the following century, the new philosophy of the scientific revolution altered notions about machines, instruments, technology and scientific knowledge. 'Philosophical machines' became a fertile focus for collective study and discourse within the new scientific societies (and, later, as means of teaching science to affluent audiences). The air pump was an important seventeenth-century example, allowing investigation of the properties of air, vacuum, respiration and life itself. Experiments employing it guided philosophical discussion of the period. Other devices rising in interest during the next century included electrical machines to accumulate charges and generate sparks and optical apparatus to investigate visual perception. Manually inventive researchers like Robert Hooke were naturally advantaged in this new environment. So, too, were commercial instrument-makers, who could now supply not just clocks, maritime and surveying instruments for navigators and land-surveyors, but research and teaching instruments to natural philosophers, too. Instrument making, in the hands of an innovative artisan, could extend science itself. As Robert Boyle noted in *The Usefulness of Experimental Philosophy* (1671), 'now we have several shops that furnish not only our own virtuosi, but those of foreign countries'.

Because of the new emphasis on empirical learning through mechanical curiosities and sense-extending instruments, natural philosophy could also be practised by an artisan, not merely someone who had immersed himself for years in reading and contemplating. Artisanal knowledge was taught by apprenticeship, and in antiquity these skills had been the domain of slaves or otherwise uneducated people. Interpretation of the surviving Greek writings had defended a hierarchy in which the intellectual developments of natural philosophy were distinguished from applied arts such as mathematics. And, with distinct cultures of recording and passing on knowledge, academic learning and craft skills had been separated. By contrast, the new philosophy highlighted the importance of empirical learning through manual expertise. It had a social consequence, too, opening up opportunities for those without book-learning or a privileged background.

Some historians have argued that the application of knowledge promoted by Bacon was especially amenable to Protestant, and particularly Methodist, theology of the eighteenth century. Having a strong ethic of valuing learning and productive work, it has been suggested that Methodists were particularly eager to relate the new philosophical ideals to business by inventing and applying new kinds of machine. Investigation of social context, in fact, has been a major thrust of the history of science since the late twentieth century, as discussed in chapter 7.

Seven decades after Bacon's death, the Royal Society focused on inventions as well as on scientific discovery through mechanical curiosities. One of the most pressing practical problems was how to pump water out of mines, a matter of considerable economic importance in the late seventeenth century. One solution, demonstrated to the Society in 1699, was the 'fire engine', or fire-powered steam engine, designed by Thomas Savery (1650–1715). Although it was scarcely practical for pumping water from deep Cornish mines, more practical types of steam engine followed. In 1712 Thomas Newcomen (1664–1729), probably adapting research begun by members of the Royal Society, invented a viable 'atmospheric engine'. His invention was responsible for the widespread application of steam engines through Europe over the next half century.

The inventions of Savery and Newcomen developed promiscuously alongside scientific investigations, but the intermingled processes can be more clearly seen in the work of James Watt (1736–1819) from the 1760s. Watt, a young instrument maker at the University of Glasgow tasked with repairing a demonstration Newcomen engine, collaborated with natural philosopher Joseph Black (1728–1799). Black's concept of latent heat, conceived later but separately by artisan Watt, allowed a more efficient steam engine design. With some initial funding from Black, Watt constructed a full scale model. He formed a partnership with Matthew Boulton (1728–1809), and Watt steam engines were sold from late 1770s for pumping water from mines. Combining engineering innovation with scientific principles, Boulton and Watt developed further refinements to expand their markets. Among them were design patents for converting the natural reciprocating motion of the pump into rotational motion, suited for weaving and milling; a speed governor (a particular example of what 150 years later would be known as a servomechanism); and an indicator mechanism for producing a graph of pressure versus volume, an important analytical and operational tool.

The steam engine, truly the motive force of the Industrial Revolution, illustrates the close affinity between technical innovation and scientific knowledge. For some of these entrepreneurs and artisans, the links were further strengthened by social mixing. The Lunar Society, an informal club and learned gathering active in Birmingham during the late eighteenth and early nineteenth centuries, included Boulton, Watt, chemist Joseph Priestley (1733–1804) and industrialist Josiah Wedgewood (1730–1795) among its members. In the wider circle of participants and correspondents were Americans Benjamin Franklin (1706–1790), Thomas Jefferson (1743–1826) and an assortment of other inventors, engineers, industrialists, politicians and men of science.

Such activities did not represent the one-way application of scientific knowledge to industry, however. The Lunar Society exemplified an entangled collection of interests having no sustained hierarchy. Through the eighteenth and nineteenth centuries in particular, the interplay of skills, information and useful application remained complex.

An illustration of how artisanal skills were often an equal or even leading partner in scientific change is the work of James Joule (1818–1889). The son of an affluent brewer, Joule took an active part in managing the business into his thirties while developing an interest in science as a pastime. His scientific studies in electricity and heat were nevertheless meticulous, and informed by his expertise in the brewing industry. His precise investigations of the mechanical equivalent of heat, for example, were founded on the careful techniques of thermometry required by brewers to control fermentation processes. Initially perceived by members of the Royal Society as a scientific dilettante and relative outsider, Joule later collaborated with William Thomson (Lord Kelvin).

Artisans such as Watt and Joule were the driving force of new, useful knowledge in the eighteenth and nineteenth centuries. Their skills made possible precise observations of new phenomena – opening territory that had not been explored by previous natural philosophers. Invention itself became important not just economically, but also intellectually and socially. Technology became a higher-status occupation and important factor in cultural change.

The Enlightenment

The scientific revolution encouraged the idea that the natural world could be better understood – perhaps even completely understood – by the application of rational methods of investigation. Over the following three hundred years, this notion spread gradually through Western societies. One of the first clear extensions of rational approaches to new domains was the movement (and period) known as 'the Enlightenment'. In its own way, the Enlightenment represents a revolutionary application of a scientific world view to the human sphere. While often excluded from histories of natural science, the Enlightenment can be interpreted as a key element in the emergence of the social sciences, as well as the evolving relationship between science and religion.

Influencing the thought of many, but not all, European scholars during the eighteenth century, this intellectual movement advocated the application of reason to all aspects of human existence. Just as Copernican observations and explanations of the solar system had challenged Aristotelian authority, so too did Enlightenment scholars challenge received authority about human affairs. Their arguments carried philosophy from examinations of the natural world to new domains: politics, morals and social institutions. In his essay *What is Enlightenment?* (1784), Immanuel Kant (1724–1804) defined it as 'man's emergence from self imposed immaturity', caused 'not by a lack of intelligence, but a lack of determination and courage to think without the direction of another'. The social changes resulting from the ongoing Industrial Revolution – many of them destabilizing and unsettling – further encouraged these attempts to rationally investigate and plan human affairs. The chemist Joseph Priestley related science directly to society. Writing in *Experiments and Observations in Different Types of Air* (1775), he argued that 'the rapid process of knowledge' had the power to 'end error and prejudice ... in this enlightened age', and warned that 'the English hierarchy, if there be anything unsound in its constitution has equal reason to tremble before an air pump, or an electrical machine'. The effects of such thought were profound and wide-ranging. The optimistic intellectual methods and social aims of Enlightenment ideas have been closely associated with science in wider culture, and continue to influence Western societies today.

As with the scientific revolution, the Enlightenment reveals no clear beginning and end points or wholly coherent thread on close examination. Viewed from a distance of two hundred years, the characteristics of the Enlightenment can be identified and described with some clarity. (This vision becomes even clearer when contrasted with very recent trends concerning science and society, as discussed in chapters 6 and 7.) Its key figures did not construct an agreed set of shared beliefs, but encouraged a general attitude of questioning based on rationality, and confidence in human abilities to comprehend and improve their existence. This scepticism about received opinion encouraged a current of profound change, though. Just as the scientific revolution had reconstructed notions of the natural world, the Enlightenment encouraged a reshaping of society founded on new principles of ethics and new forms of social organization.

Kant argued, for example, that judging appropriate human conduct should be a matter of reason, not religious authority or social traditions. His system of ethics provided a reasoned framework that could be critiqued and modified by others. Voltaire (1694–1778), a generation older than Kant, concluded in the same vein that experience and reason could improve the human condition. His prolific writings illustrated his opposition to the dogmatism of organized religions but also his criticism of atheism, which he felt would lead to a loss of morality. Voltaire promoted the ideals of intellectual liberty and argued against the tyranny of governments. Thomas Paine (1737–1809), an Englishman participating in the American Revolution in 1776, later wrote *The Rights of Man*, a discussion of Enlightenment ideals of equality and liberty, and a defence of the French Revolution.

The outbreak of revolution in France in 1789 was not triggered directly by Enlightenment thought: violence and civil unrest did not sit easily with rational enquiry and planning. During the first few years of the revolution, political change was prioritized over natural philosophy. However, the abstract ideals led to radical, if brief, reorganization of French society in the new republic. Calendars and time-keeping – weeks, days, hours and minutes – were reorganized into a decimal system. Weights and measures were converted from the mediaeval French units to a new metric system. The new standard of length, the *metre*, 'rationally' based on the circumference of the earth rather than human scales such as the foot and inch, embodied the ideals of reason over tradition, and objectivity over subjectivity.

Contributors to the Scottish Enlightenment emphasized the importance of pragmatic solutions, rejection of authority and importance of reason even more than their continental European counterparts did. David Hume, for example, sought to develop a 'science of man', applying scientific study to preceding human cultures. His definition of reliable knowledge, based on factors such as experience, evidence and causation, were important in developing a philosophically grounded scientific method. Adam Smith's *Wealth of Nations* developed a theory of economics that was quickly adopted by the British government. Given the intellectual momentum provided by such thinkers, Scottish science continued to be disproportionately influential through the nineteenth century.

The Enlightenment is sometimes identified as drawing to a close during the time of Napoleon in the early nineteenth century. The optimistically radical ideals of the French Revolution had, it seemed, been brought down to earth by a pragmatic dictator. Nevertheless, Enlightenment ideals, closely pinned to evolving scientific practice, were further developed during the nineteenth century.

The Darwinian revolution

The waves of change carried by scientific, industrial and social revolutions were more easily traced in retrospect. Acting over a time scale of decades and involving numerous historical actors, their influences shaped every aspect of human culture. A more focused scientific revolution, but also having enduring cultural consequences today, was associated with Charles Darwin's publication *On the Origin of Species* in 1859. As with other periods of remarkable change that we have already encountered, this shift was more obvious in retrospect and can be linked with wider social changes beyond the *Origin*.

Darwin (1809–1882) had studied medicine at Edinburgh University, intending to follow a career in medicine like his father, but became interested in natural history there. Seeking a more secure future, his father enrolled him at Cambridge

University to study theology. Upon completing his studies in 1831, however, Darwin secured an unpaid position as naturalist and gentleman's companion on the *HMS Beagle*, which was tasked with charting the coastline of South America. Over the next five years, Darwin collected animal, plant and fossil specimens and studied the geology of the continent and surroundings, taking extensive notes. He was particularly interested, for instance, in finding evidence to decide between two contemporary theories of geological change. Georges Cuvier (1769–1832) had proposed *catastrophism*, in which brief but violent events play the dominant role in large-scale transitions such as extinctions and mountain formation. The explanation, reliant on exceptional and rare serial catastrophes, was consistent with Biblical interpretation. By contrast, Charles Lyell (1797–1895) claimed that *uniformitarianism* fitted the geological evidence more closely. According to this gradualist account, the physical processes responsible for geological formations have operated unchanged throughout the earth's history. However, such gradual change required the Earth to be much older than suggested by contemporary interpretations of Biblical chronologies.

The letters recording Darwin's findings made his reputation as a naturalist even before his return to England in 1836. His large botanical, geological, ornithological and fossil collections were studied carefully by his contemporaries, and his *Voyage of the Beagle* (1839) brought further renown. Benefiting from his father's investments, Darwin was able to live as a gentleman scientist.

A keen observer and analyst, during his travels Darwin had noted variations between species of birds and tortoises on the Galapagos Islands, and mused about the stability of species. From 1837, he recorded his developing ideas about the transmutation of species in private notebooks, supporting them with experiments on plants and work in animal husbandry. Over the following twenty years, he refined his arguments for the evolution of species and discussed his ideas with a handful of his contemporaries. When Darwin learned during the late 1850s of similar work by naturalist Alfred Russel Wallace (1823–1913), however, he completed his carefully argued *Origin of Species*. In it, he provided the evidence for 'common descent', i.e. the evolution of species from a common ancestor, rather than one or more separate creations of species as then commonly believed. Darwin argued that the driver for this process of change was *natural selection*, in which the individuals most suited to survival would transmit their successful characteristics to their offspring. In environments that culled a significant fraction of a species through competition for resources, the inherited attributes of the population would therefore gradually alter.

Criticizing contentious claims

The response to Darwin's ideas illustrates the many forms of criticism that scientific claims can undergo. Indeed, examinations of debate and contentions have increasingly attracted historians of science. Darwin's theory provoked intense interest among the public and scholars alike. Popular interest had been captured in the decade before *The Origin* by an anonymously published popular book, *Vestiges of the Natural History of Creation* (1844). Its author, journalist Robert Chambers, had postulated that all aspects of the natural world, from the solar system to rocks to living organisms, had transmuted from more primitive forms. While only loosely analogous to Darwin's rigorously argued thesis and suggesting no mechanism for the process of change, it had encouraged wide discussion of the gradual alteration of species.

Indeed, the idea was not novel, and other scholars had considered the alteration of species much earlier. For instance

Darwin's own grandfather, the intellectual Erasmus Darwin (1731–1802), had argued that all life arose from a common ancestor. In his book *Zoönomia* (1794–1796), he suggested that new parts and propensities were acquired by a species as a result of 'irritations, sensations, volitions and associations', and so could introduce improvements 'by its own inherent activity'. Among the factors affecting the nature of life he speculated that competition and sexual selection might be important, notions which his grandson grounded in specific mechanisms and evidence.

Jean-Baptiste Lamarck (1744–1829), writing at the beginning of the nineteenth century, devised explanations of the alterations of species by a natural propensity to increase order and complexity, and by their environmental experiences. Uniquely, he claimed that the more frequent use of a bodily characteristic would result in it being inherited in offspring. These acquired characteristics would result in species adapting to their local environment.

Despite such earlier ideas, Darwinism was nevertheless challenged by a variety of critics. As Darwin himself realized, there was no convincing explanation in his time for how hereditary traits could be communicated intact to the next generation. It was thought that offspring shared equally in the characteristics of their parents, so biological characteristics should be diluted from generation to generation. Only at the turn of the twentieth century was the work of Gregor Mendel rediscovered, which provided strong evidence for laws of genetics.

One of the most damning criticisms of the period concerned the *philosophical materialism* claimed to be inherent in Darwin's theory. Materialism holds that the natural world can be fully understood in terms of physical matter, and that no immaterial or supernatural factors are required. Darwinism, it was argued, replaced God by an unguided natural process. Such criticism had been aimed by an earlier generation at the mechanical philosophy, which in the seventeenth century provided a mathematical explanation of the solar system in terms of the concept of gravity. Darwinism carried such thinking to a new stage, however: such natural 'laws' threatened Biblical interpretations and, by implication, the special creation of humans themselves by the conscious intent of a creator.

Philosophical materialism: the study of explanations of nature and existence wholly in terms of material and physical properties and laws.

Many of the criticisms had been anticipated by Darwin himself, who had buttressed his arguments with copious evidence and discussion. His chronic ill-health, however, left public rebuttals to others who championed his concepts. The most influential of these during the Victorian period was biologist Thomas Huxley (1825–1895). Known as 'Darwin's Bulldog', Huxley engaged in a widely publicized and influential debate in 1860 with the Bishop of Oxford, Samuel Wilberforce. Huxley, a skilled anatomist, also countered arguments that there were aspects of the human brain that were qualitatively unlike other species. The hippocampus, he demonstrated, was present in the brains not only of humans but of other primates such as the gorilla, too.

Gorillas fascinated the Victorians. The first Western accounts of live gorillas were published only during the 1850s, and specimens were first exhibited in Europe and North America a decade later. Displayed in cages or even dressed in children's clothes, primates were disturbingly human-like. Not surprisingly, popular accounts of Darwinism conflated it with this side-show interest.

Darwinism, particularly in Britain, threatened the established social conventions, as suggested by a cartoon in the periodical *Punch* (Figure 6). If species evolve, it seemed to imply, then are

Figure 6 *Punch* cartoon, 'The lion of the season' (*Punch*, 25 May 1861)

not all men – and apes, too – cousins? And, if so, what about the comfortably fixed divisions (at least for the affluent privileged classes) according to class or race?

Too often Darwinism has been presented as a clash of ideologies that split society into opposed camps. The reality is more complex. The popular response in Darwinism can be compared with the indiscretions of President Bill Clinton while in office. His publicized affair generated moral outrage for some, provided a target of jokes for others, and indifference from yet others. A hint of these plural responses towards Darwin is contained in *Princess Ida*, an 1884 comic opera by William Gilbert and Arthur Sullivan. While focusing on the ridiculousness of higher education for women – an idea having both religious, political and intellectual overtones at the time – it also poked fun at Darwinian notions, showing that apes, like men, were unchangeable despite their desires for improvement:

> A Lady fair, of lineage high,
> Was loved by an Ape, in the days gone by.
> The Maid was radiant as the sun,
> The Ape was a most unsightly one –
> So it would not do –
> His scheme fell through,
> For the Maid, when his love took formal shape,
> Expressed such terror
> At his monstrous error,
> That he stammered an apology and made his 'scape,
> The picture of a disconcerted Ape.
> ...
> He bought white ties, and he bought dress suits,
> He crammed his feet into bright tight boots –
> And to start in life on a brand-new plan,
> He christened himself Darwinian Man!
> But it would not do,
> The scheme fell through –
> For the Maiden fair, whom the monkey craved,
> Was a radiant Being,
> With brain far-seeing –
> While Darwinian Man, though well-behaved,
> At best is only a monkey shaved!

But Darwinism was interpreted selectively to suit more serious social convictions, too. Charles Darwin's cousin, Francis Galton, conceived *eugenics* in 1883. According to Galton's interpretation, Darwinism implied that, in the absence of intense competition, species would be weakened or diluted. Galton's solution was to ensure the selective propagation of those individuals deemed most fit. Conveniently for those often in a position of power, the 'most fit' were conventionally defined as the affluent and healthy, or those whose forebears had been so. How, it was simplistically argued, could they have achieved their social position otherwise? By contrast, contemporary definitions of the hereditarily unfit included criminals, the poor and the disabled. Never supported by Darwin himself, eugenics attracted prominent supporters among intellectuals and politicians, and by the turn of the twentieth century was a popular concept embodied in a growing number of laws and selection practices. Eugenics could take 'positive' forms (encouraging reproduction by those categorized as genetically advantaged) and 'negative' forms (discouraging reproduction by individuals or groups that were deemed to be genetically disadvantaged). It provided a 'scientific' gloss for policies implementing, for example, racial segregation, limiting national immigration and compulsory sterilization in mental hospitals. In tune with these popular understandings, it also was used as an argument by early proponents of birth control and genetic screening.

In a similar way, *social Darwinism* extended Darwin's concepts of biological selection to the social sphere. It suggested that competition not just among individuals, but also at the level of groups or entire societies, produces social evolution. Although Darwin mused about the evolution of 'social instincts' and 'moral sentiments' that might favour one nation over another, the idea reflected contemporary understandings amenable to Victorian attitudes and circulating since the Enlightenment. Societies, it was argued, progressed through stages of improvement. The British Empire of the late nineteenth century, some claimed, was the culmination not merely of social organization – customs, laws and cultural refinements – but also a biological selection. Thus the British could be argued to be the rightful heirs of empire based on superiority of their traditions, institutions and biology. Since the mid-1940s, both eugenics and social Darwinism have been attributed to the more explicit and equally derided racial notions of the Nazi regime. The crude notions of norms and objectively definable progress that underlay eugenics and social Darwinism have increasingly been challenged by historians and other scholars during the twentieth century, as discussed in chapter 6.

Darwinism achieved a political status in the USA after the First World War. There, a few southern and largely rural states passed laws that prohibited the teaching of the theory of evolution in schools. The laws were motivated by literal interpretations of the Bible. These concerned the age of the earth – which was conventionally calculated from Biblical chronologies to be a few thousand years old – and the special creation of mankind 'in the image of God'. The best-known outcome was the Tennessee trial of schoolteacher John Scopes in 1926. Scopes was found guilty of having taught evolution, but the press coverage of the trial, both national and international, mocked the prosecution team and solidly supported Darwinism

Darwinism continues to provoke debate from certain publics today, notably some American fundamentalist Christians, as discussed in chapter 4. Within decades of the publication of *The Origin* the debate moved, however, from the scientific sphere to the religious domain, where it has been mired in small pockets of ardent criticism ever since. The 'Darwinian revolution' depends on perspective, then. The rapid expansion of intellectual terrain for practising men of science and philosophers has slowed to a manageable pace as evolutionary biology has become embedded in experimental practice and continuing theoretical development. By contrast, its social and religious implications have flared up irregularly. For its non-scientist critics, the revolution is ongoing and contested.

Knowing a revolution when you see one

It is worth making a diversion at this point to explore the notion of revolution a bit further as a postscript to the chapter. If you prefer to continue the chronological survey, skip on to the next chapter, and return to this section after chapter 6.

If the shift towards Darwinism can be questioned by some audiences today, it suggests that identifying such transitions has much to do with the judgments of social groups as well as to intellectual persuasiveness. The American physicist and historian Thomas Kuhn (discussed further in chapter 7) skirted these dimensions when he published the first analysis of historical episodes of science in terms of revolutions. While he limited his attention to scientific opinion (instead of attending to non-scientists such as those that keep alive the debate about Darwinian evolution), Kuhn explored elements that had not previously been considered important in the replacement of one way of thinking by another.

Kuhn's analysis of the scientific revolution explained it as the exchange of one *paradigm* for another. According to his admittedly imprecise definition, a paradigm is an explanatory model that comprises scientific concepts, methods, facts and assumptions. Kuhn argued that, instead of altering ways of thought gradually and progressively, a dramatic paradigm shift in scientific beliefs could occur with unexpected suddenness. The shift from one intellectual framework to another could result, for example, by the accumulation of awkward facts that fit poorly with an existing theory. When the body of ill-fitting facts was too weighty to reconcile with the orthodox worldview, a new framework would be developed to replace it, and consensus would tip towards the new position. If the new world view were radical enough, the old one might even become literally incomprehensible, a process that Kuhn called *incommensurability*.

This rather satisfying interpretation carries with it considerable ambiguity. Kuhn recognized an essential tension in judging whether a revolution was in progress or not: were the awkward facts merely inaccurate observations, or were they crucial evidence of the failure of existing models? For the historical actors, this would be a difficult judgment call. What factors lie behind assessments of 'awkward data' or an 'elegant theory'? Even murkier dimensions were to be found in the social interactions that underlie scientific verdicts. The quality and relevance of awkward facts might be disputed, based on how the evidence was acquired and by whom. Was the investigator a member of the discipline and known to be competent? What affiliation of interests supported the orthodox model, or could be rallied around the new paradigm? Changes in scientific belief, it was apparent, rely on more than compelling evidence: we need to pay attention, for example, to the cultural factors that make it seem compelling.

A good illustration of Kuhn's approach, along with the difficulties of historical interpretation, is the case of Alfred Wegener (1880–1930). Wegener, a German meteorologist, postulated the idea of 'continental drift' during the early twentieth century. He hypothesized that the continents were mobile, and cited evidence accumulated from diverse sources. Wegener was struck by the complementary coastlines of South America and Africa, and noted that the other continents could be rearranged into a reasonably close-fitting configuration (dubbed 'Pangaea'). He also found unusual similarity in the geological and fossil record of those close-packed regions, even though they are today separated by oceans. In effect, Wegener identified 'awkward facts', and marshalled them to challenge orthodox geology. The conventional explanation – that coastline shapes were coincidental and that fossil similarity could be explained by land bridges – appeared much more plausible to contemporary geologists, though. Wegener's explanations for continental drift (centred on attributing continental movement to centrifugal force of the rotating earth and to tidal forces from the sun and the moon) were derided both by geologists and physicists (although gaining some support from life scientists). For a generation thereafter, Wegener's ideas were unpopular.

After the Second World War, however, geological measurements of seafloor spreading and earthquake zones, combined with Wegener's more superficial evidence, revived interest in continental movement. The new theory of *plate tectonics* provided a satisfying explanation of continental movement to geologists. Continents did not plough through the earth's crust, as Wegener claimed; they floated on upwelling magma. These syrupy convection currents, not astronomical forces, drove the slow collisions of tectonic plates to generate earthquakes and form mountains. Within a single generation, a dramatic shift in scientific consensus took place.

Was this a revolution in Kuhn's sense? Maybe, although opinions vary about precisely how it fits. Many geologists today identify Wegener as the founder of a new paradigm, even though his 'awkward facts' and concept of continental drift differed in most scientific respects from those of plate tectonics. Wegener's revolution failed for intellectual and social reasons. His evidence was interpreted by his contemporaries as handpicked and circumstantial, too light to tip the balance towards a radical new model. His theoretical explanations were demonstrably incorrect according to modern physics and geology. And, as a meteorologist dabbling in the discipline of geologists, he was judged to be both presumptuous and ill-informed. By contrast, the post-war flip to plate tectonics has characteristics much closer to the large-scale shifts we have encountered in this chapter. Significantly, it relied on the painstaking design and operation of varied scientific instruments to acquire data that compelled new theoretical explanations. Wegener, then, teeters between fame and irrelevance depending on just what we mean by 'paradigm'. Historical interpretation can be a cruel business!

4

Spreading a seductive idea

This chapter focuses on five themes that rose to relevance during the nineteenth century, but had earlier origins: the rising appeal of science for the wider public; the changing relationship between science and religion; the growing confidence and enthusiasm for mathematical representations; the creation of 'scientific' medicine; and, the rise of government-managed science. If the roughly chronological flow and organizing themes raise unanswered questions for you, please be patient. This historical survey will be interrogated and challenged in chapters 7 and 8.

A democratic science?

Science as an educational or amusing diversion became a more common activity of the prosperous classes during the seventeenth century and even more so during the eighteenth. A trend toward popular uptake of scientific ideas continued to spread through the next two centuries. Practical engagement with science involved not just an intellectual or economic motive, but enjoyment as a popular pastime as well.

During the late eighteenth century, public demonstrations of scientific phenomena proved increasingly popular. The philosophical machines first studied by members of scientific societies proved a popular way of demonstrating the taming of nature by science, and so for entertaining, inspiring and educating small audiences. For instance one of them, the 'Electrical Kiss', was often demonstrated as a parlour trick for private audiences. A couple would touch lips after the operator of a hand-cranked electrical generator charged up the female participant standing on an insulating platform. The resulting static shock was simultaneously thrilling, embarrassing and awe-inspiring: a memorable illustration of the power of scientific practice! Other popular demonstration phenomena that raised public curiosity about science – often to much larger audiences – included laughing gas (first produced by Joseph Priestley [1733–1804] in 1772) and *animal magnetism*, popularized by Franz Mesmer (1734–1815) also during the 1770s.

Natural history, focusing on the classification of plants and animals, proved a more accessible study. Such collections had become popular with medical men during the sixteenth century, and during that period compilers of natural history catalogues increasingly sought to apply principles of logical division. The Swedish naturalist Carl Linnaeus (1707–1778), in his *Systema Naturae* (1735), devised a system of classification and nomenclature which encouraged the search for new specimens and their suitable ranking. Natural history collections became part of the Enlightenment project to develop systems of classification for the natural world, and so affluent collectors could simultaneously both pursue a hobby and contribute to an intellectual goal. Botanical collecting, for example, became a popular pastime and vocation for a variety of audiences ranging from young ladies to clerics. Other subjects for collections included insects, minerals, fossils, shells, taxidermic specimens and, sometimes within the same collections, antiquities such as coins and ethnological artefacts such as headgear and cutting implements. All could be catalogued, illustrated, inter-compared, differentiated and classified in the attempt to systematize nature and even human societies.

Affluent collectors of fossils, demonstration gadgets, archaeological artefacts, stuffed animals and 'freaks of nature' built up private 'cabinets of curiosities'. While these were intended originally to impress visitors with the wonder and mystery of nature (and simultaneously the wealth and urbanity of the collector in possessing nature), such accumulations often became the basis of nineteenth-century museum collections. Such publicly exhibited collections were often intended as a form of rational education to assist 'the diffusion of a taste for Natural Science amongst the working classes', as noted by the Natural History Society of Northumberland in 1836. An even more portable and easily spread variant was the natural history print. As detailed engravings (often hand-coloured) which could be bound and published in books or displayed in frames, such prints encouraged close examination of nature in a more amenable context.

A more organized form of popular exploration of science was introduced at around this time, though. The first literary and philosophical societies were founded as discussion clubs on intellectual topics but specifically excluding religion and politics. They also frequently were repositories for private natural history collections. Such 'Lit & Phil' societies were founded in many English towns, particularly in the north, from the 1780s to 1830s. These reflected the increasing interest in science by industrialists, and shifted the centre of mass of scientific life from the largest urban centres. These new groupings brought together the manufacturers, physicians, entrepreneurs and intellectuals discussed in chapter 3. The Manchester Lit & Phil included, at various times, chemist John Dalton, James Joule and Joseph Whitworth, promoter of standardization in engineering.

While these societies opened scientific discourse to relatively affluent engineers and scientifically minded men, the spread of scientific education and discussion to working men (and often women) was promoted by so-called *mechanics' institutes*. The promoters of the institutes were often those who frequented the

Lit & Phil Societies. Philanthropists and industrialists, among the first funders of such organizations, argued that this would provide a positive evening activity for employees and create a more skilled (and perhaps motivated and malleable) work force. Free lectures on the arts and sciences had been provided in Glasgow from the turn of the nineteenth century, but the first mechanics' institute focusing on technical subjects was founded there in 1821. The idea spread rapidly over the next decade. Within a half century some seven hundred institutes were active in towns around the world, including not just urban centres such as London, Melbourne, New York and Montreal but also small towns including Wednesbury, Hedon and Dumfries. The curricula of the institutes were meant to provide deep, if simplified, scientific knowledge rather than mere technical skills. Scientific facts and laws were taught as grounding for the occupational skills of the adult students.

Science was also disseminated less prescriptively to increasingly wide audiences through periodicals. Until the nineteenth century, scientific journals had been written for, and made accessible to, practising men of science affiliated with societies. Most of the popular publications, though, were generalist periodicals covering a broad range of topics. They communicated articles ranging from descriptive accounts (e.g. 'Observations of the glow worm', in *Wesleyan Methodist Magazine*, 1823), science education (e.g. 'Lessons in botany' in *Englishwoman's Domestic Magazine*, 1853) and even current scientific debates (e.g. 'the craniological controversy' in *Edinburgh Magazine*, 1817). Later in the century popular journals such as *Scientific American* (starting as a one-page newsletter in 1845) and *Popular Science* (1872) chronicled the latest inventions and innovations for American popular audiences. These had evolved from short-lived publications of local mechanical engineering clubs and reflected popular interest in the rapid expansion of railways, telegraphy and, later in the century, electric lighting and telephony. Written to be intelligible to broadly educated audiences, they promoted a democratic tendency for scientific knowledge and linked it with the modern world.

This expansion was fuelled, too, by improvements in publishing technology. Steam presses and then mechanical typesetting equipment, introduced early in the century, increased production rate and lowered cost, bringing newspapers and journals to mass audiences for the first time. Even more importantly, illustrations became more common in books and periodicals as copper engraving was replaced increasingly by wood and steel engraving and lithography. Collectively, such articles melded the findings of science with a sense of wonder and a growing Victorian confidence in progress.

Science and religion: liberation or conjoined twins?

Previous sections have hinted at, but skirted over, the theological dimensions of scientific ideas. The next three sections address them more directly. Historians of science, theologians and others have examined their interactions, often suggesting a confrontational relationship between the methods of science and other forms of knowledge, particularly theology. More recent scholarship has suggested, though, that the relationships have been complex and variable. Historical re-evaluations have identified episodes of coexistence, attempts at harmonizing their claims, and even religious encouragement of scientific activities.

As we have seen, early Christians absorbed much of the Greek science available to them to create a world view consistent with their interpretation of scripture. Similarly, Islamic scholars, although adopting a more active strategy of experimenting empirically, did not often identify a problem of reconciling scriptural interpretations with observations of the natural world. One example of conflict, though, was the work of the Egyptian Muslim scholar al-Ghazâlî (1058–1111). In his book *The Incoherence of the Philosophers*, he criticized previous generations of Islamic philosophers who had built on the writings of Aristotle and Plato, and argued that their studies were inconsistent with Islamic faith. Even so, al-Ghazâlî's criticisms concerned metaphysics rather than the Greek sciences of astronomy, physics and logic: his objections centred on philosophers' rejections of the impossibility of disrupting causality (in effect, denying the existence of miraculous interventions – an issue for some Christian theologians, too) and of providing proof of the existence of God as creator of the universe. It has been suggested by some scholars that this assertion of faith in the direct influence of God discouraged subsequent scientific explorations of regular laws. On the other hand, medicine (particularly anatomy) and astronomy continued to flourish in the Islamic world over the following two centuries.

For Judaism, the other major Western religion, overt conflict between science and religion is also difficult to discern. Jewish commentators even at the time of Darwin argued, for example, that the rabbinical tradition encouraged freedom of thought in both scientific and religious views. Without an authoritative single voice, the interpretations of the Torah supported dialogue and reason in determining the make-up of the physical world.

Putting humankind in its place

While Islam and Judaism were influential in Europe, where science expanded most dramatically, Christian dominance of government, law and religion carried a more influential authority. Three historical episodes are commonly cited to explore the relationship between Christianity and science, all of them in the period during or after the scientific revolution. The first involved the controversy surrounding theories of the universe, particularly whether it was earth-centred (geocentric) or sun-centred (heliocentric). The second was the rise of the 'mechanical philosophy', according to which nature could be understood via universal laws. And the third was Darwinism. Of course, these three episodes, all well before the twentieth century, do not exhaust the possibilities for discussion. They do, though, introduce and illustrate recurring points that still evoke discussion today.

When Galileo published his support of Copernicanism (see chapter 3) in *Dialogue Concerning the Two Chief World Systems* (1632), he attracted the ire of the Catholic Church. His writings eventually were banned and he was required to recant their claims. The episode is less clear-cut than commonly portrayed, though. It is also important to note that criticism did not begin there: Martin Luther (1483–1546) had ridiculed Copernicus's ideas of a moving earth as early as 1539. As recorded in a collection of his conversations in *Tischreden* (*Table Talks*, 1566), he reflected the contemporary scientific and religious consensus:

> There is talk of a new astrologer who wants to prove that the earth moves and goes around instead of the sky, the sun, the moon, just as if somebody were moving in a carriage or ship might hold that he was sitting still and at rest while the earth and the trees walked and moved .. The fool wants to turn the whole art of astronomy upside-down. However, as Holy Scripture tells us, so did Joshua bid the sun to stand still and not the earth.

Many of the leaders of the Protestant reformation were equally dismissive.

For Galileo nearly a century later, the run-up to censorship was gradual. He had first faced an enquiry (*inquisito*, or inquisition) after publishing *Sidereus Nuncius* in 1610, two decades before his *Dialogue*. As a stubborn and forthright scholar, he had invited opposition from notable critics such as the Archbishop of Florence. Galileo's telescopic observations and interpretations seemed to challenge conventional Biblical understandings. The Bible appeared to state that the sun moves (*Joshua* Chapter 10, verse 13), that the earth is motionless (*Psalm* 104) and that the heavens are arranged in a circle around the earth (*Isaiah* Chapter 40, verse 22). The heliocentric universe devised by Copernicus denied all these Biblical interpretations.

More seditiously, an earth that moved around the sun meant that humans were not the centre of the universe and, by implication, not the centre of God's plan. This notion appeared to unbalance views that had been established since the time of the Greeks. The 'Great Chain of Being' had illustrated a natural order. Humans had dominion over lower species. Lower down the chain were simple animals, plants and finally inanimate rock. Christians revised the plan only slightly: God, then angels, held positions above humans in the hierarchy. A geocentric universe mapped this hierarchy neatly: God and heaven lay above celestial bodies and the earth at the centre of creation. The anthropocentric (human-centred) arrangement of the universe seemed to confirm the cosmology developed by Aristotle and Ptolemy as well as Biblical accounts of the place and purpose of humans in God's creation. A heliocentric universe, by contrast, relegated humans to the status of God's afterthought.

The confrontation between the geocentric and heliocentric theories raised other theological dimensions, too. The development of the 'mechanical philosophy' was founded on a kind of faith of its own. This scientific faith, though, was continuously extendable. Its supporters claimed that the characteristics of the natural world were regular and discoverable. Miracles, if they took place at all, had to be infrequent. Otherwise, there was little point in trying to discover regularities that could be bypassed at God's whim. A natural world dominated by miracles would encourage passivity and acceptance of the nature of things, rather than active exploration. In fact, this was the unwanted advice lobbed by one of Galileo's critics, who quoted on the pulpit Biblical admonitions about the futility and impiety of examining God's work too closely ('Ye men of Galilee, why stand ye gazing up into heaven?' [Apostles chapter 1 verse 11]). Second, mechanical philosophers trusted that these regularities could be detected and explained. Galileo's success in describing acceleration by a mathematical expression was one example of their growing confidence. But the culmination was Isaac Newton's theory of gravity. His gravity was not merely a mathematical description, or law. It also conveyed a relationship between the heavens and the earth. Newton's law pointedly explained the motion of the moon and planets by the same law followed by Galileo's earthly objects.

Newton was a relative latecomer among the challengers of philosophical and religious orthodoxy, but his mathematics offered new perspectives for discussion. A generation earlier, the French philosopher René Descartes (1596–1650) had attracted ardent supporters to the mechanical philosophy with his new approach. Importantly, he was able to define principles and concepts that were seductive in the same way that Aristotle had been to previous generations. They appeared able to provide a satisfying answer to every phenomenon. This convincingly plugged a hole opened by earlier critics of Aristotelianism. Copernicus had proposed a mathematically elegant explanation of celestial movements, but was at a loss to explain earth-based physics. Galileo had extended mathematical explanations to a few key earth-bound experiments, but did not generalize further. Descartes, though, applied the same spirit to make sweeping generalizations that many of his contemporaries found liberating and optimistic.

On the other hand, just like Aristotle's explanations, many of Descartes' answers were criticized as being inadequately mathematical. In fact, Newton sought to define his own work in relationship to his predecessor: while Descartes had entitled his chief work *Principia Philosophiae* (*Principles of Philosophy*, 1644),

Newton emphasized the mathematical foundations of his own by crafting the title *Philosophiae Naturalis Principia Mathematica* (*Mathematical Principles of Natural Philosophy*, 1687). Both Descartes and Newton can be called 'mechanical philosophers' because they broadly applied mechanical notions to explain the natural world, but their concepts were wildly different. Descartes postulated the existence of an undetectable substance, the *plenum*. As Aristotle had argued, the notion of empty space is unintuitive; bodies would surely move quickly to fill it up. The plenum, an infinitely subtle substance filling all space, could explain the orbiting of planets around the sun. Descartes argued that the sun created a vortex of this material, which carried the planets around it much as an emptying bathtub makes a toy boat circle the drain. By contrast, Newton rejected the plenum and claimed that the space between the planets was completely empty of material. How else could objects move without slowing down? To explain the motions of the solar system, he invoked his own imperceptible entity: gravity.

By the time of Newton's death in 1727, Aristotelian ideas had been trounced, but the Cartesian and Newtonian explanations of the natural universe vied for followers. As Descartes's countryman Voltaire wrote that year, national allegiances played a role in defining truth:

> A Frenchman who arrives in London finds a great alteration in philosophy, as in other things. He left the world full; he finds it empty. At Paris you see the universe composed of vortices of subtile matter; at London we see nothing of the kind ... Among you Cartesians all is done by impulsion; with the Newtonians it is done by an attraction of which we know the cause no better.

In the same way, the mathematics developed in the *Principia* – Newton's 'method of fluxions' – was challenged on the Continent by another formulation ('calculus') devised by his

German contemporary, Gottfried Leibniz. The mechanical philosophy, then, offered distinct options not just for scientific practice, but for interpreting the underlying order, too.

Creation and constancy: a new place for God

Between the conceptual differences of the new philosophers, the role of God was re-evaluated. Both Descartes and Newton sought to illustrate and explain God through the mechanical philosophy. For Descartes, God was the creative force and perfect being. Theology was also an important part of Newton's scholarly life; he devoted at least as much time to biblical analysis as to physical science. Convinced that biblical texts had been corrupted by early copyists, he collected and inter-compared accounts. (He devoted this same meticulous attention to assessing current knowledge of alchemy, another lifelong and equally private interest.) He sought unsuccessfully to discover messages encoded in his corrected texts. Newton opposed the Anglican views of his day, and resisted taking holy orders, an expectation of Cambridge scholars. As an anti-Trinitarian, for example, Newton questioned the belief in three distinct and equal entities (Father, Son and Holy Spirit) and opposed the worship of Jesus as God.

Newton was uncomfortably aware that his work and religious ideas challenged conventional understandings. He opposed the growing idea of a mere clockwork universe. The observed regularities and mathematical relationships, he argued, required – and even demonstrated the existence of – God. In effect, the role of God had been shifted: no longer seen as responsible for moment-by-moment control of the natural world, God was increasingly understood as the conceiver and inventor of a complex but decipherable system that could thereafter run itself. Yet Newton himself was uneasy about the nature of gravitation, and mused about the degree of God's management of it. In his *Principia*, he argued that not only did the cosmos 'proceed from the counsel and dominion of an Intelligent Being' but that 'This Being governs all things, not as the soul of the world, but as Lord overall'.

Most of his supporters were unaware of Newton's private religious views and developed a variety of positions on the role of God ranging from a distant creator to an omniscient governor of the universe. Proponents of Newtonianism during the eighteenth century were, for example, frequently supporters of *deism*. (Benjamin Franklin was a prominent example.) Deism is the theological view of God as an inventive mechanic who does not interfere with the subsequent operation of the universe. In its most abstract form, deism may not attempt to describe the characteristics of such a non-interventionist creator, or even that the universe is identical with God (a variant known as *pandeism*). Some additional qualities attached to deism. For example, deists criticized the notion of 'received knowledge', e.g. revelation from sacred scriptures. The expanding confidence in mechanistic explanations also suggested to deists that life forms, like the inanimate world, could be described by regularities of cause and effect. If so, it raised religious and philosophical questions about our ability to individually choose rather than having options determined by natural laws.

While exploring sometimes dangerously unorthodox religious ideas, most natural philosophers of the eighteenth and early nineteenth centuries supported some concept of a higher being or creator. This notion fitted the context of the times, and was well expressed by so-called *natural theology*. William Paley (1743–1805) wrote *Natural Theology; or, Evidences of the Existence and Attributes of the Deity* (1802). The book introduced a popular form of the 'argument from design' that has circulated since. The argument (not a deductive argument in the logical sense, but one that attempts to persuade by analogy) is sometimes known as the 'Watchmaker analogy'. If we were to find a pocket-watch during a walk, he argued, we would know instantly that it was not an accidental creation, but instead that it had been conceived by a designer. In the same way, the intricacies of the natural world compel us to believe in a designer; it is inconceivable that such elaborate designs could possibly have arisen otherwise. This claim has been revived in the modern world: so-called 'intelligent design' asserts that this baffling complexity cannot have arisen through the random processes of Darwinian evolution. While promoted principally by fundamentalist Christian activist groups to influence educational options, intelligent design nevertheless avoids direct claims about the attributes or identity of the designer.

Examples illustrating the argument from design, such as the complexities of the eye or of gestation, were frequently cited during the early nineteenth century to explain the reasonableness of natural theology. Even in an age when the regularities of physics suggested understandable laws, biology remained mysteriously complex; both fields demanded a fundamental cause or origin. Natural theology foregrounded a long-standing concern that was being gradually set aside by natural philosophers: what is the *origin* and *purpose* of the natural world?

A popular series of books, *The Bridgewater Treatises*, set out during the 1830s to illustrate 'the Power, Wisdom, and Goodness of God, as manifested in the Creation'. Volumes by eight philosophers and scientists appeared successively, seeking to show evidence of God's design in nature. William Whewell, for example, wrote *Astronomy and General Physics Considered with Reference to Natural Theology*, and concluded that the existence of God was readily evident. A ninth fragmentary text by Charles Babbage, inventor of the first mechanical programmable computer or 'difference engine', sought to portray God as the programmer of an elaborately complex natural world.

As a student of theology at Cambridge University, Charles Darwin read Paley's *Natural Theology* and found it persuasive. His later voyage on the *Beagle*, though, suggested to him that the variations and distribution of different species were not good evidence for intentional design by a creator. Darwinism, introduced in chapter 3, challenged the Watchmaker analogy (and has proved equally successful against 'intelligent design' in legal challenges in American courts). It provided an unsettling alternative to the comfortable reassurances of natural theology. Darwinism's scientific arguments had a theological side-effect. This was close in effect to what the heliocentric theory had done to astronomical understandings. Copernicus had shown that humans were not the centre of the physical universe. Darwin's theory demonstrated the humans were not separately created outside nature; they were a species within it, subject to the same environmental pressures and mutations as other living things. Through the advance of Darwin's ideas, anthropocentrism and its religious support were further undermined.

Each of these conceptual controversies revealed long-standing tensions between theology and rationalism. Deism and natural theology were two examples of accommodation. Another interesting juxtaposition was the 'religion of humanity' devised by Auguste Comte (1798–1857) to replace conventional religions. According to Comte's categorization, religious knowledge represents an early and imperfect stage in human development. In the absence of evidence to the contrary, it argues, discussion of the existence or non-existence of God is meaningless. He recognized, however, the social power of religion. Comte's work, like those of a number of his French contemporaries, must be understood in the context of Enlightenment thinking, particularly following the French Revolution of 1789–1799. During the early nineteenth century the questioning of ideologies and institutions became freer. Comte himself argued that the church had long imposed constraints on society.

His secular religion was modelled closely on Catholicism, replacing key concepts with equivalents that he drew from science. Feast days, for example, commemorated scientists, philosophers and poets; church services were replaced by worship in Temples of Humanity, and a mother and child, representing Order and Progress respectively, bore a strong resemblance to the Virgin Mary and Jesus.

The term applied to this form of unrefined faith in the powers and benefits of science is *scientism*. While such confidence in the authority of science grew steadily through the nineteenth century, today the term scientism is most often used in a pejorative sense to represent dogmatic and excessive trust along the lines that Comte constructed. In fact, some recent critics of the scientific enterprise have attempted to argue that scientific tenets themselves may be founded on faith rather than empirical evidence. Such challenges – from the social sciences as well as religious critics – are discussed further in chapter 7.

Scientism: the conviction that the methods, criteria and conclusions of science provide the only legitimate foundations of knowledge.

Atheism – the denial of the existence of God – is not an automatic consequence of scientific, positivist or even scientistic views. Scientism concludes that scientific methodology provides the only reliable knowledge, and that theological questions are meaningless. Practising scientists during, and since, the nineteenth century more generally subscribed to some form of scientific method but held a variety of religious beliefs (including atheism). A vocal modern proponent linking atheism with good scientific practice is the evolutionary biologist Richard Dawkins. It is fair to say that his stance has been popularized even more by its opposition to the influential claims of American fundamentalist Christians. While generally supporting

Dawkins' arguments, a larger number of scientists have intellectually separated private religious convictions from their science, arguing that theological ideas are largely irrelevant to their work.

The cachet of numbers

The foundation for Comte's secular religion was his creation of *positivism*, a system that amalgamated science, philosophy, history, politics and social studies around an idealized view of scientific knowledge. While the grand organization of his Religion of Humanity was a failure, his philosophical writings made a significant impact. Comte based his notions of a hierarchy of sciences on a simple framework. He ranked forms of knowledge according to their explanatory success. At the top of the pyramid was mathematics itself. Mathematics, he argued, is universal – verifiable by any intelligent mind – and exact. And astronomy, the oldest of the observational sciences, owed its success to its mathematical precision, through which celestial regularities were successively discovered. Comte related the power of a scientific domain to its age and its degree of mathematization. Lowest on the scale, but having the highest hopes of improvement, was his new subject of sociology.

Others were reaching similar conclusions. Adolphe Quetelet (1796–1864) analyzed aspects of human variation such as height and weight, observing that such characteristics of populations followed a bell-shaped curve. As an astronomer and founder of the Royal Observatory of Belgium in 1828, he noted the similarity between these human variations and the distribution of astronomical measurements. Quetelet proposed that such regularities followed 'social laws' akin to the 'natural laws' determined by natural philosophers. Indeed, he entitled his principal work *La Physique Social* (*Social Physics*, 1869). For instance, Quetelet applied statistical methods to define the 'body-mass index' to determine normal weight, and the concept of *l'homme moyen* (the average man) based on the most common values of human attributes. He extended these techniques to correlate social behaviours – notably criminality – with factors such as poverty, alcohol consumption, gender, education and even climate.

Living organisms, too, were increasingly submitted to the scientific gaze and mathematical description. The field of *psychophysics* is a good example of the expansion of these methods into the human realm. Coined by Gustav Fechner (1801–1887) in 1860, the term refers to the study of human perceptions by the methods of physics. His aim was to transform psychology, understood at the time as a philosophical study, into a positive science. Seeking a mathematical relationship between a physical stimulus and the resulting perception, he determined a law: the intensity of a sensation increases arithmetically (i.e. linearly) if the stimulus increases geometrically (i.e. exponentially). All five senses – vision, hearing, touch, taste and smell – became the basis of quantitative measurements.

Fechner's research was extended by scientists such as Wilhelm Wundt (1832–1920), who founded the first psychology laboratory at the University of Leipzig in 1879 and the first journal for the subject two years later. Wundt sought to understand the human mind as the composition of distinct faculties (although this was not a new idea: it had been claim of phrenologists from the turn of the century). He introduced techniques that were widely adopted, such as determining the threshold at which a sensation was just detectable, or by balancing two stimuli to achieve the same sensation. These procedures proved moderately repeatable on different human subjects, and gave confidence that reliable quantitative regularities could be determined. Wundt's examination of isolated characteristics, and his employment of controlled experiments to build a theory of sense perception, distinguished his approach from earlier work and proved to be an enduring feature of psychological research.

Within the physical sciences, mathematical foundations were now solidly embedded. Since Newton's *Principia*, mathematical description had thoroughly colonized astronomy and physics and was making inroads into chemistry. Engineering knowledge, traditionally acquired through apprenticeship and experience, was increasingly allied to mathematical physics. The words of one of the most prominent inventor-physicists of the century, William Thomson (Lord Kelvin, 1824–1907) summed up the consensus:

> I often say that when you can measure what you are speaking about, and express it in numbers, you know something about it; but when you cannot express it in numbers, your knowledge is of a meagre and unsatisfactory kind; it may be the beginning of knowledge, but you have scarcely, in your thoughts, advanced to the stage of science, whatever the matter may be.

Although these words were part of an 1883 lecture on electrical units, they have been frequently emblazoned on the engineering buildings of twentieth-century universities.

Hermann von Helmholtz (1821–1894) straddled the fields of physics and physiology; Wilhelm Wundt had been his student. Helmholtz integrated research that previously had been the subjects of separate disciplines. In studying acoustics, for example, he developed instruments to generate sound and combined them with experiments on human subjects to determine perceptions of pitch and volume. On a much grander scale, his widely admired *Handbook of Physiological Optics* (1867) expounded experiments and theories of depth and perception, colour vision and visual physiology. The approach adopted by Helmholtz and his colleagues – subjecting all aspects of the natural and living world to the methodology of physics – turned its back on the ideals of an earlier generation of his countrymen, the followers of *naturphilosophie* (see chapter 7). As historians

Peter Bowler and Iwan Rhys Morus aptly summarize it, 'where their predecessors had wanted to show that the universe could be treated like a living organism, the new generation of physiologists wanted to show that living organisms could be treated like machines'.

As Helmholtz's work hints, human vision proved to be a fertile source of new phenomena during the nineteenth century, and increasingly was subjected to a mathematical approach. Inventions by men of science included the kaleidoscope, stereoscope and motion picture apparatus such as the kinescope. Each provided insights founded on mathematical analysis, and each became available as a mass-produced entertainment. As discussed in chapter 3, industrial applications of science became a feature of nineteenth-century innovation. Through manufacturing industry, these optical devices quickly became popular with large audiences.

Less visible to the public but no less significant economically, the quantification of light became important at the end of the century. The development of electric lighting generated keen competition with the gas lighting industry established over the previous decades. This commercial rivalry was supported by consulting scientists, who measured and judged the lighting systems according to quantitative criteria. How efficient was the generation of light, and how did it depend on its fuel, age and other parameters? How stable was the light produced, and how was it distributed? Just as importantly, how could town managers monitor the quality of gas and electricity supplies, and manufacturers ensure consistency in the light bulbs they produced? The answer was provided by an increasingly standardized process of judging light intensity using groups of human observers. In effect, human perception was rendered machine-like to yield objective measurements. The eventual replacement of human observers by measuring instruments finally occurred during the 1930s, despite severe difficulties in defining the 'average observer' according to strict mathematical relationships. As H. D. Murray, a scientist summarizing the field, later noted, 'Simplification of complex situations is a feature of all physical measurement and it has been nowhere more extensively applied than in subduing colour to the requirements of measurement'. Applying mathematics to living organisms was acknowledged to require them to be yoked and subdued to their applications.

Thus the momentum of an increasing scientific fashion carried quantification to other disciplines and to wider publics. Established sciences such as astronomy became the model for chemistry and sociology. Engineering adopted the calculations of physics, and promoted further expansion of mathematical tools such as calculus. Inventors and their clients adopted mathematical discourse to compare products. The seduction of numerical description compelled would-be sciences to adopt the tools of physics. In the following century, this cachet for the quantitative swept over popular culture, too.

What made medicine scientific?

This growing confidence in numbers had an impact in fields of knowledge that historically had been distinct. As we have seen, the appeal of quantification successively seduced physicists, engineers and philosophically oriented psychologists. The work of Comte, Quetelet and others applied mathematical methods to studies of society. Notions of typical humans – and of their deviations, too – were quantified. And during the nineteenth century, numerical measurement also began to seduce medicine.

Treating medicine as distinct from science may seem, at first glance, unwarranted. The justification for doing so is that 'scientific medicine' represented the expansion of one thread of an ancient and evolving body of knowledge. This development was recent in historical perspective, becoming evident only during the late nineteenth century. And, in professional terms, the history of medicine has traditionally been treated on its own terms by its own specialists. As discussed in chapters 6 and 7, the history of science has been a broad church that only recently has amalgamated historians of technology, sociologists, philosophers and, not least, historians of medicine.

What, then, was 'scientific medicine'? Certainly a close link had long existed between natural philosophy, biology and medicine as suggested by Hippocrates' four *humors* (chapter 2). Such approaches to medicine were as systematic and popular as Aristotle's physics and cosmology were. And, in the same way, these medical systems gradually were challenged by evolving practices of observation, dissection and experiment during the period of the scientific revolution. Established traditions were increasingly tested, challenged and augmented or replaced.

Another way, then, of identifying the distinct approach of scientific medicine is by examining changing medical practices. Consider, for example, the procedures adopted at the Crichton Royal Hospital, an innovative Scottish mental institution founded in 1839 (which, ironically, my own university campus now occupies). The surviving records of case notes indicate that on admission the earliest patients would undergo minimal medical examination but a diagnosis by interview and evaluation of reference letters. A description, sometimes lengthy, described the demeanour, habits and mannerisms of the patient, but measurements were limited to pulse and respiration.

During this period, the hospital environment itself represented the principal means of therapy. The hospital incorporated facilities graded by social class, a cathedral-like church for worship, a farm for therapeutic physical labour, an art studio and gardens for relaxing promenades (indeed, the buildings themselves were sited to create pleasing views of the surrounding hills and to obstruct view of the nearby town from which many of the patients came). Drug treatments were limited mainly to 'narcotics and laxatives'.

Until the late 1870s, the succeeding administrator frequently advocated recourse to restraint and seclusion and, for 'the better class patients', calm intellectual pursuits. During the 1880s the 'Segregate System' was introduced, providing separate buildings for different diagnostic categories of patient. By the turn of the century, blood pressure was recorded as well as pulse and respiration, and a limited number of blood tests were introduced.

From the mid-1930s doctors there implemented laboratory research (neurology, pathology and psychiatry) and sought more active intervention in psychiatric illness by implementing various shock therapies such as prolonged narcosis (extended sleep induced by narcotics), electro-convulsive therapy, insulin therapy (the induction of severe lowering of blood sugar) and pre-frontal leucotomy. From the 1950s therapies were also supplemented by newly available psychotropic drugs.

The Crichton site represented changes in the profession of medicine, its institutions, organizing concepts, practices and technologies. The sketch of its changing practices suggests an uncontested advance, and indeed histories of medicine have often promoted this simple view. It is important to recognize, though, that the influx of scientific methods into medical practice was resisted by many doctors, who protested that their expertise was being displaced or denied, and by many patients, who balked at an increasingly impersonalized relationship with the doctor.

Medical diagnosis traditionally had relied on holistic evaluation based on examination and questioning of the patient. By the late nineteenth century, this method of gathering information was notably different from, for instance, new experimental methods in psychology. Some medical practitioners began to argue that the traditional method was subjective, depending on an experienced but possibly biased medical practitioner. If so, measurements might provide a more objective evaluation of deviations from norms. This seeking of scientific objectivity was a distinct change of direction for medical practice.

The tools of the medical profession broadened as the profession itself was extended. The work of Florence Nightingale (1820–1910) during the Crimean War illustrated the virtues of a professional class of nurses trained in meticulous observation and employing statistical methods to evaluate and compare treatment regimes. From the turn of the century, temperature measurement became a nursing speciality. The thermometer (or fever) nurse became expert at literally charting the progress of an illness and adapting therapies to suit.

Medical institutions themselves adopted features that were becoming common in the academic, engineering and commercial spheres. Organization of patients by category of illness, as at the Crichton, was proven by examples such as Nightingale's, and mirrored engineering practices promoted by 'efficiency engineers' such as Frederick W. Taylor from the late nineteenth century. (The seduction of Taylor's 'scientific management' was to influence twentieth-century organizations dramatically.) Differentiation of hospital wards and the division of labour among medical staff became increasingly refined through the twentieth century. Wards under the eye of a single senior generalist doctor at the beginning of the century were being transformed into more specialised units under a hierarchy of practitioners after the First World War.

The increasing adoption of laboratory tests owed much to biological research, which was also adopting experimental methods. Such research expanded along three distinct fronts from the late nineteenth century, each seeded by scientific methods. First, public health research had demonstrated from the 1840s that infectious disease could be sourced to environmental conditions such as contaminated water supplies (see the following section for more on this). Studies of slum conditions revealed further correlations between health and sanitation, income and diet. Reductions in infectious diseases such as cholera, typhus, diphtheria, malaria and smallpox were attributed to the measures to improve such conditions. In short, disease during the Victorian period was found to have a significant *social* component.

Second, eugenicists (see chapter 3) argued for *hereditary* causes for disease. They promoted isolation hospitals to prevent the transmission of certain illnesses to the general population and to subsequent generations. And third, chemists and biochemists were studying *biological* causes for disease. By the turn of the century, compelling evidence for a germ theory of disease had been acquired by researchers such as French chemist Louis Pasteur (1822–1895) and German physician Robert Koch (1843–1910). Thus three versions of scientific medicine appeared within a couple of generations.

A pharmaceutical industry also grew rapidly at the end of the nineteenth century, incorporating research to discover new medications. An increasing array of biochemical products targeted the destruction of microorganisms responsible for particular illnesses. During the twentieth century, successful vaccines were devised to combat, for example, diphtheria (1923), poliomyelitis (1954), mumps (1967) and rubella (1970). Also in the period after the First World War, the development of hormonal treatments such as insulin for diabetes transformed a growing number of chronic illnesses. The discovery of correlations between dietary deficiencies eradicated others: vitamin A for night blindness, vitamin B for pellagra and beriberi, vitamin C for scurvy, vitamin D for rickets.

Medical instrumentation also linked closely with developments in physics and engineering. The adoption of x-rays for diagnosis of broken bones and embedded bullets was rapid after its invention in 1896 (see chapter 5 for more). Technologies of health and illness represented cultural shifts not just for medical personnel, but for patients, too. Significantly, the delusions reported by some of the Crichton patients frequently made reference to the technologies of the day: persecution via railways, galvanism and the telegraph up to the 1870s, by the telephone and electricity during the 1880s, and wireless communication and vacuum cleaners from the 1920s.

Together, these features of scientific medicine transformed the treatment of illness. Lawrence J. Henderson (1878–1942), a prominent American medical scientist, claimed that from the turn of the century, 'A random patient, with a random disease, consulting a doctor chosen at random had, for the first time in the history of mankind, a better than fifty-fifty chance of profiting from the encounter'.

Pride and panics: science and the State

Another important theme of the practice of science during the nineteenth century – but with earlier beginnings – was the role of the State. The involvement of governments in what had been conceived as a philosophical activity during the Enlightenment was motivated by discrete events. This sporadic involvement, in turn, led to more overtly political dimensions to science during the twentieth century.

As illustrated by Francis Bacon's writings in the seventeenth century, scientific activities had long been understood as having a relationship to power. That recognition by specific rulers was nevertheless implemented considerably later. Scientific societies, appearing from the mid-seventeenth century, brought together local concentrations of scholars but, in 1666 – a generation after Bacon's death – the Académie Française des Sciences was founded in Paris. The Académie was supported by Louis XIV for the national encouragement of French science. By the end of the century, it incorporated formal rules, permanent headquarters (in the Louvre) and a growing collection of paid scholars and assistants. Its seventy members founded a botanical garden, commissioned research and mounted large-scale expeditions to make astronomical observations. Such information gathering was understood to serve national application (for example, celestial observation promoted accurate navigation during a period when exploration and colonial holdings were highly active) and to be important for the reputation of France. Just as importantly for this discussion, they were available to provide advice to the State.

By contrast, the relationship between the Royal Society and the State was at arm's length. Months after assuming the throne at the Restoration of the monarchy in 1660, the new English king, Charles II, paid lip-service to the French by becoming patron to the Royal Society, but his patronage of this 'college of physico-mathematical experimental learning' did not extend to funds for paid scholars. Science remained largely outside the government's remit, and a laissez-faire attitude prevailed. Why, indeed, should the State become involved in this activity, that was partly a pastime, partly a fashion, partly a profit-making enterprise? The attention remained episodic, in fact, and was punctuated by what could be characterized as periods of pride and panic.

In 1851 Queen Victoria's consort, Prince Albert, sponsored the 'The Great Exhibition of the Works of Industry of all Nations'. As an affluent enthusiast, he was typical of the rising Victorian popular interest in science and engineering. In the Crystal Palace, a mammoth and innovative structure of glass and iron speedily assembled in London's Hyde Park, the exhibition showcased British innovation and vaunted the country's pioneering successes of the Industrial Revolution. Nearly one-third of a mile long and a hundred feet high at its centre, the Crystal Palace enclosed 14,000 exhibits that allowed comparison of the technological products of the British Empire and other countries. The exhibition proved wildly popular and was followed by similar exhibitions in other countries: New York (1853), Paris (1855), for the Fine Arts and then, in 1867, a shocker: the Exposition Universelle in Paris. The exposition, with national pavilions for the first time, exhibited some 50,000 displays from forty-one countries. British observers were jolted by the evidence of other nations' innovations and large investment in science and technology, and feared for the consequences for international trade.

Other exhibitions followed: in Paris about every eleven years, and in Prague, Antwerp, Chicago, Melbourne and Buffalo by the end of the century. Each caused more worry, and even panic in certain quarters, in Britain. One outcome was in promoting higher education. Some one thousand men studied science in British universities in the 1860s, but this rose fivefold within twenty-five years, and then by a further factor of five before the First World War. The funding of science departments and academic specialists was a key factor. In 1883, for example, the Scottish universities of Glasgow, Aberdeen and Edinburgh opened faculties of science. Even more directly, government legislation created an Inspectorate of School Science in the Science and Art Department at South Kensington, a product of the Royal Exhibition. The aim was to ensure that school curricula would teach science to produce a commercially competitive generation.

A separate motivation for government involvement in science was health and safety, which also led to control via inspectorates. Three large-scale manufacturing industries led to serious environmental pollution during the nineteenth century. Gas lighting, invented in the 1810s, led to large-scale manufacture of gas from coal. From mid-century, synthetic dye production also promoted large-scale chemical plants. And the cotton industry became increasingly reliant on coal-burning steam engines for mills and the production of soda (alkali) to bleach cloth. A by-product of the LeBlanc soda process was the airborne release of enormous quantities of hydrochloric acid vapour. In 1850, for example, the English town of Widnes burned over two million tons of coal for its soda plant, or about one thousand tons per head of population. Around such towns, vegetation and wildlife – as well as the local population – were endangered. The Alkali Act (1863) appointed five inspectors to curb emissions from LeBlanc plants. These first government chemists were able to visit factories, but were unable to impose changes for another forty years.

An even more pressing problem through the nineteenth century was cholera. It first became a world-wide problem from the 1820s, after the British East India Company conquered the Bengal. Beginning there, an epidemic spread to Britain and America in 1831. There were three subsequent pandemics (1849, 1853 and 1866), with some 115,000 deaths in the UK and about 200,000 in America. In 1854 John Snow, doctor to Queen Victoria, correlated cholera outbreaks with tainted local water supplies. This methodical deduction led to a near-consensus about the scientific explanation of a socio-medical problem, and led the government to support new standards of water cleanliness. The Public Health Act (1875) was a bold intervention that relied on trust in scientific evidence.

So government support in Britain was curiously expressed. The Victorian State engagement with science was largely through inspectors of schools, chemical plants, gas and water supplies. All were scientifically trained government employees tasked with the responsibility to assure quality.

Later in the century, and led by Germany and France, such interventions moved from preventative to supportive science through the creation of 'national laboratories'. The French government set up the Bureau International des Poids et Mesures (1875) to regulate measurements and, with them, to expand exports based on metric units. The German government created the Physikalisch-Technische Reichsanstalt (1887), a

State institute in experimental physics to promote, as industrialist Werner Siemens (1816–1892) put it, the 'advancement of science and, thereby, also the technology closely bound to it'. Britain's National Physical Laboratory (1899) and America's National Bureau of Standards (1901) followed it. All three focused on supporting industry through research to provide national competitive advantage. Typical tasks included the design of ship's hulls, the standardization of electrical products, and the determination of physical constants. At the opening of the twentieth century, then, State science was an established and expanding principle.

5 Twentieth-century turns

This chapter – and, more indirectly, the final two – focuses on the past century. Coverage of the most recent hundred years or so, in fact, makes up nearly half the book. These chapters are intended to foreground my intention declared at the outset: to survey not merely the history of science, but also the 'history of history of science', i.e. the corresponding discipline. At this point it is enough to note that the twentieth century holds great interest for popular audiences and for professional historians of science today. Science was transformed during the last century as it explored new domains and responded to new cultural pressures. Expanding in activity and cultural relevance, it became more overtly and recognizably technological, commercial, militaristic and political. It affected not just specialist communities but the lives of everyone. The peculiar trajectory of science finally attained a global significance.

Fin de siècle science

Artificial as they are, the beginnings and ends of centuries are traditionally periods of cultural reflection and renewal. The beginning of the twentieth century was typical, and illustrates the extent to which science had become embedded within Western culture.

The term *fin de siècle* has been used since the 1890s to refer to this period expressed by distinctive themes in art and literature. It has become a short-hand for a cultural mood: a jaded world-weariness that combined sophistication and decadence. In a sense, everything wholly novel had been explored, and new experiences had to be sought in nuances or new directions. It was nevertheless a time of great creativity in the arts, which were establishing new forms of visual representation (e.g. French Impressionism and Art Nouveau), literary expression (e.g. the *stream-of-consciousness* of James Joyce) and theme (e.g. the eroticism of Aubrey Beardsley).

This cultural mood extended to the practice of science. The end of an era was commonly recognized by scientists such as Albert Michelson (1852–1931), who argued that twentieth-century physics would be devoted to pursuing the next decimal point in precision measurement to reveal ever more subtle phenomena. The nineteenth century had suggested that physics, at least, was well explored and explained. Newton's mathematical representations of mechanics had been developed and systematized with great success in the intervening two centuries by physicists such as Pierre Simon Laplace (1749–1827). Thermodynamics, organized by the newly invented concepts of energy and its conservation, had been constructed in mid-century by a close merging of experimental investigations and mathematical explanations. James Clerk Maxwell (1831–1879), during the 1870s, provided a similarly powerful theoretical understanding of electricity and magnetism that explained all the experimental observations of Michael Faraday (1791–1867) and others – but significantly it *followed* those experimental investigations. The methods underlying these understandings were also rendered routine: by the end of the century, national standards laboratories were being established or mooted in Germany, Britain and America. This expansion of scientific knowledge, and potential end of new horizons, recalled Bacon's writings of three centuries earlier suggesting that active science might be an activity requiring only a few generations of effort.

Despite the sanguine forecasts of older physicists at the turn of the century, however, the recognition of wholly new phenomena soon generated waves of investigation and application. The extended episode, while a remarkable series of events in the history of science, illustrates the typical cultural effects of science in the new century. In the space of barely a decade, a class of new phenomena was discovered, explained and applied; new professional groupings and commercial markets emerged; and, significantly, the public engagement expanded swiftly.

The new phenomena appeared gradually from research centring on electricity. In 1886 the German physicist Heinrich Hertz (1857–1894) demonstrated the generation of electromagnetic (or 'Hertzian') waves by electrical sparks. These were undetectable except via their effects on other spark apparatus. During the next decade a series of other emission phenomena were reported. Figure 7 shows varieties of 'Crookes tubes', named after the English scientist William Crookes (1832–1919), who first employed them to study a new phenomenon. (The same Crookes, incidentally, who was sometime President of the Royal Society, the Institute of Electrical Engineers, the Society of Chemical Industry and the Society for Psychical Research; the frontiers of these phenomena-rich fields fascinated him.) The unexplained attribute of Crookes tubes was a faint blue glow that surrounded one of the electrodes when a voltage was applied. These so-called *cathode rays*, which caused certain materials to fluoresce, were studied by physicists and by inventors studying electric lighting. Physicist J. J. Thomson (1856–1940) used Crookes tubes to identify the rays as streams of electrically charged 'corpuscles', later dubbed *electrons*. From this family of devices came the cathode-ray tube (CRT) used for television and computer displays, and also the *vacuum tubes* or *thermionic valves* used for the first generation of electronics over the following fifty years.

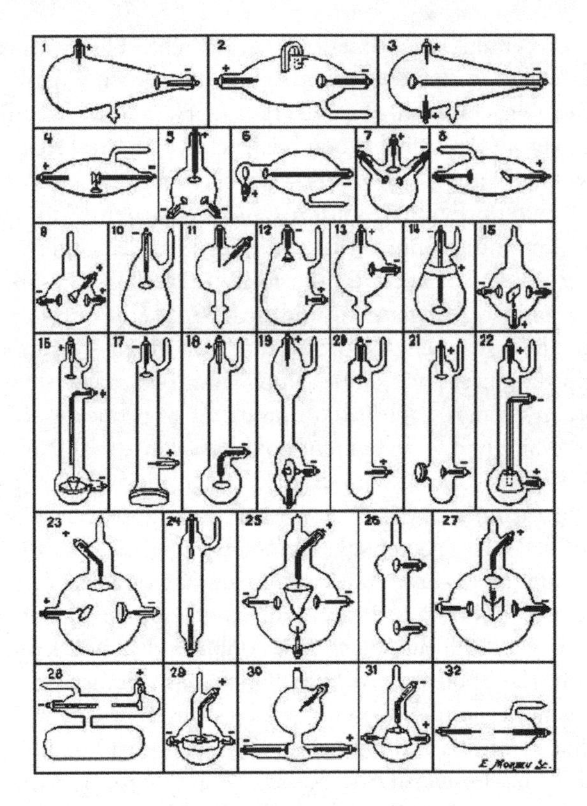

Figure 7 Varieties of Crookes tube c.1900

While using such tubes in 1895, Wilhelm Roentgen (1845–1923), the director of a German physics institute, discovered another type of emission. His 'x-rays' – producing shadow images of bone in living subjects, and soon claimed to be of value in killing germs and treating illness – were a phenomenon with no convincing explanation for over a decade, and yet had clear potential for practical application.

Investigators sought to relate the new rays to known properties. The faint glow of the cathode ray seemed similar to fluorescence of certain rocks exposed to sunlight. Henri Becquerel (1852–1908) in France found that such rocks (particularly uranium salts) emitted invisible rays that could expose film in sealed packages, just like Roentgen's x-rays. Pierre (1859–1906) and Marie Curie (1867–1934), also in France, were able to isolate two previously unknown constituents in granite (polonium and radium) that produced similar effects. New Zealand physicist Ernest Rutherford (1871–1937), working in Canada and later England, found that Becquerel's rays could be deviated by a magnetic field, and that cathode rays responded in the opposite direction. These 'alpha' and 'beta' rays, along with Roentgen's x rays, each had a unique set of properties of interest to seemingly diverse branches of physics. In 1900 Paul Villard (1860–1934), working with uranium, identified another form of emission ('gamma rays'). These were not stopped by a layer of paper (like alpha rays) or several meters of air (like cathode rays) or by bone (like x-rays), but could penetrate dozens of meters of concrete. Gamma rays, it seemed, could not be influenced by electromagnetic disturbances, but could be detected from rocks and from the open sky. And finally, again in France and beginning from experiments with x-rays, physicist René-Prosper Blondlot identified another form of emission that he dubbed 'N-rays' after his Université de Nancy. N-rays appeared in certain materials exposed to sunlight and, like Hertzian waves, were found to have a subtle effect on light sources. Italian investigators suggested that N-rays were emitted by the human body, and could be used to monitor blood flow.

This last variety of 'alphabet ray' provides a twist to the story, though. Although some fourteen investigators, mainly in France, reported successful experiments with N-rays, evidential support for their existence declined and, by 1904, they were deemed to be figments of the investigators' imaginations. Given the rapid pace of new discovery during the decade, and the diverse origins of these new effects in rocks, sky, fluorescence, electromagnetism and sparks, it seems Blondlot and his colleagues were misled by the subtlety of their observational techniques. This is not to say that one research group failed to meet prevailing standards of scientific practice, but that the scientific practices themselves were on the edge of practicability and unguided by established expectations. Even so, Blondlot and some colleagues continued to support the reality of their discoveries for years thereafter. The episode has been an interesting battle-ground between scientists and historians contesting how science and pseudo-science can be distinguished.

In brief, then, international research by a generation of physicists demonstrated an exciting spectrum of new phenomena. The scientific implications were radical – so radical, that an argument can made that the turn of the twentieth century was faster paced and more revolutionary than the present day. So, is science accelerating or not? As the period illustrates, scientific innovation does not always correlate with research budgets and numbers of workers.

New frameworks for new phenomena

Beyond the wholly new phenomena surrounding 'alphabet rays', physics was also being challenged by a new extension of Newton's laws that united mechanics with electromagnetism: this 'special theory of relativity' introduced in 1905 by Albert Einstein (1879–1905) made a growing impact over the next twenty years, especially when extended by his 'general theory of relativity' in 1916. With the 'special theory', Einstein straightforwardly pursued the consequences of explaining electromagnetism in a consistent manner. This required jettisoning some long-established ideas, notably the notion of a preferred 'frame of reference': Einstein argued that the laws of physics should be identical whether we are moving (e.g. on a train) or not. Indeed, he showed that it is impossible even to determine velocity in an absolute sense.

This provided a satisfying explanation (although Einstein probably was unaware of it) for the failure to detect such motion experimentally. During the late 1880s, the American physicist Albert Michelson (1852–1931) and chemist Edward Morley (1838–1923) had sought to confirm the existence of the *luminiferous aether*. (This is not the same as the classical aether discussed by Aristotle.) For late Victorians, the aether seemed to be a hopeful basis for the unification of physics. Nearly a century of optical experiments – particularly by French investigators such as Augustin Fresnel (1788–1827), François Arago (1786–1853), Leon Foucault (1819–1868) and Hippolyte Fizeau (1819–1896) – had constructed an understanding of light as a wave. All waves required a medium such as water, air or other fluid to propagate through. The luminiferous aether was that medium, and was assumed to fill all space. From the frequency of light waves, it was possible to calculate the mechanical characteristics of the aether: higher frequency requires a 'stiffer' elastic medium – and yet, unintuitively, one that does not impede the motion of objects through it.

So confident were many physicists of the necessity of the aether that Lord Kelvin was able to lecture in 1884 that 'we know the luminiferous aether better than we know any other kind of matter in some particulars'. Kelvin, having been impressed by the analogy of smoke rings and how stable they could be in air, proposed 'aether vortices' to explain atoms and their chemical combinations. (There is a certain similarity to Descartes' *plenum* here.) James Clerk Maxwell, the Scottish physicist who provided a mathematical synthesis of electromagnetism, even devised a mechanical model to represent the role of the aether in its phenomena. This strong dependence on mechanical analogies was, in fact, a feature of physics in the English-speaking world at the end of the nineteenth century.

In any case, such a medium supporting light waves should be detectable by its optical effects. The Michelson-Morley experiment was intended to detect daily and annual variations in the speed of light as their apparatus travelled, along with the rotating earth, in orbit around the sun – moving alternately with, then against, the fixed aether. Michelson explained the absence of any observed effect by insulation of earth-bound experiments: aether was carried along with the earth's motion, reducing the observed effect. Einstein's Special Relativity later offered a more radical explanation (although he probably was unaware of the Michelson-Morley results): the speed of light was the same for all observers, *whether they were moving or not*. With the eventual acceptance of relativity, physicists shed the explanatory need for aether, although some of its attributes have been revived in more recent research. Even more counter-intuitively, Special Relativity implied that no body could exceed the speed of light c, and that mass and energy were equivalent (and related by the formula $E = mc^2$).

Incidentally, Michelson, like Blondlot and his faith in N-rays, never gave up his conviction that motion through the aether could be detected, and he and followers devised and conducted ever-more elaborate and sensitive optical experiments until his death. This suggests at least two things to subsequent historians: that the objective pursuit of knowledge can be tenacious enough to override social norms or, as with Blondlot, that scientific reputation plays an important role in enrolling support. Philosophers would add that the case illustrates how empirical evidence may be explained by more than one theory.

Even more controversial than Special Relativity was Einstein's General theory. General Relativity extended his notions from 'inertial' frames (objects moving at a constant velocity) to accelerated frames, including the effects of gravity. Central findings were that space and time must be considered together, and that gravitation can be understood as a distortion of space-time. Its mathematical consequences were worked out over the following decades, and included the prediction of black holes, regions of intense gravitation that even light cannot escape.

These developments eventually transformed aspects of twentieth-century physics and astronomy (where many of Einstein's predictions have observable consequences) but made little initial impact; most of the experimental tests of General Relativity became feasible only towards the end of the twentieth century. Of even more importance to physics at the turn of the century was the hypothesis of Max Planck (1858–1947) that all radiation is transferred in individual energy elements, or *quanta*. Planck's hypothesis was able to explain the spectrum of radiation of heated bodies – a puzzle of practical relevance to lighting manufacturers and to researchers at the new Physikalisch-Technische Reichsanstalt. Einstein proposed further that light itself was quantized (into packets later dubbed *photons*) to explain how light caused cathode rays to be emitted from vacuum tubes. These hypotheses led to a flurry of international research that accelerated through the century, and led to the new field of *quantum mechanics* between the two world wars. Like relativity, quantum mechanics fostered dramatic alterations to scientists' world views. Nineteenth-century debates about the nature of light – whether it was a particle or a wave – fell away: it was now decided to have the characteristics of both, depending on the experiment.

The mechanical models so popular with Victorian physicists appeared increasingly sterile on the atomic scale. At the turn of the twentieth century not only rays, but particles were becoming the focus of physics. The new alpha, beta and gamma rays appeared to be closely connected with the properties of matter.

At laboratories around the world, small groups of physicists explored the new terrain. The groups shared findings and expertise via journal publications and research students (who commonly studied at a handful of European centres and moved to new sites as their careers advanced). The New Zealand physicist Ernest Rutherford and English chemist Frederick Soddy (1877–1956), for example, explored radioactivity at McGill University in Montreal. Rutherford demonstrated that the radioactivity of different elements declined with a characteristic exponential decay and well-defined *half-life*. Soddy coined the term *isotope* to explain how the same chemical element could be measured to have different specific weights. Later, at the University of Manchester and the Cavendish Laboratory, Cambridge (where he succeeded J. J. Thomson as Director), Rutherford spawned a generation of internationally linked investigators.

Through such groups in Germany, France, Italy, Britain and America during the early twentieth century, the properties of the atom were actively explored. Its characteristics were unfamiliar at best. Thomson had imagined it as a kind of 'plum pudding', with small negative electrons embedded in a heavier positive material. Rutherford's group, by contrast, argued that their experiments explained the atom as a compact heavy nucleus surrounded by orbiting electrons, with empty space between.

Imaginative experimental work was accompanied by equally novel theorization. The Danish physicist Niels Bohr (1885–1962) first made his name by proposing a very non-mechanical model for the atom of hydrogen. He imagined it to be something like a miniature solar system, with a proton at its centre (or *nucleus*) and a smaller electron orbiting it, but required that the electron release and absorb energy in distinct amounts. Rather than spiralling gradually toward the nucleus, the electron would revolve only in fixed orbits, hopping between them when it absorbed or emitted a photon of light. His contemporaries argued that matter at such a scale could not be understood in macroscopic terms. In fact subatomic particles such as electrons and protons could only be tracked to a certain precision: either their momentum or their position could be precisely known, but not both simultaneously. This 'uncertainty principle' was first described in 1925 by German physicist Werner Heisenberg (1901–1976).

But this proliferation of new physics did not unify the subject as Victorians had hoped: a century later, relativity cosmology (superbly applicable on the astronomical scale) and quantum mechanics (equally accurate on the subatomic scale) have not been reconciled. The uncertain principle became a centrepiece for the so-called 'Copenhagen interpretation' of quantum mechanics developed at Niels Bohr's physics institute in Denmark. Einstein himself rejected the notion of uncertainty that was at the centre of contemporary interpretations of quantum mechanics, famously observing that 'God does not play dice'. He and other physicists devised thought experiments highlighting seeming inconsistencies and paradoxes in Bohr's interpretation, with some seeking detailed deterministic models of subatomic physics that dispensed with limits on certainty. Einstein himself sought a unified field theory – a means of merging gravitation with electromagnetism and other fundamental forces – unsuccessfully until his death in 1955. The reconciliation of relativity and quantum mechanics has remained an active research problem since then.

Biology, too, was revolutionized by the rediscovery in 1900 of the experiments of Gregor Mendel (1822–1884), an Austrian cleric who had studied the inheritance of traits in pea plants. During the 1860s, Darwin himself had proposed a provisional hypothesis that he recognized as inadequate and qualitative, known as *pangenesis*. According to this, the traits inherited from parents were due to microscopic components ('gemmules') shed by the various parts of their bodies, which were transmitted and combined at fertilization. According to pangenesis, parents could transmit bodily characteristics that they had developed during life, rather than merely by inheritance. For example, the loss or disuse of a body part would mean that its gemmules could not contribute to inheritance. This possibility – that a parent's experiences could influence its progeny biologically – had first been introduced by Lamarck, but was not supported by experiment. By contrast, Mendel's careful investigations, conducted at the monastery where he later served as abbot, established that biological characteristics could be explained by the transmission of 'recessive' and 'dominant' components responsible for particular features such as pea colour and surface texture. Mendel's evidence was so clear-cut, in fact, that historians have subsequently questioned whether he was the victim of *confirmation bias* (the tendency to interpret data in ways that favour a hypothesis, for example by unconsciously rejecting inconvenient samples as bad). A related idea is Pierre Duhem's notion of *theory-ladenness* – the difficulty in conceiving fresh interpretations of experiment when perception is clouded by an existing theoretical framework. His results were nevertheless readily confirmed and generalized by other investigators. For biologists of the period – particularly plant breeders – Mendel's work provided a satisfying explanation of biological inheritance and a detailed mechanism for Darwin's theory of evolution. Over the next two decades and with the significant application of statistics to the field, the subject of evolutionary biology was rapidly established. Fertilized by the new subject of genetics, Darwin's foundations were buttressed and increasingly mathematized.

The beginning of the century was thus a time of tremendous upheaval and considerable uncertainty in science. New phenomena, entities and concepts inspired researchers, motivated industrialists and excited a growing popular audience.

Commercial science

What Roentgen called the 'x-light' provides a good illustration of the expanding economic, social and cultural dimensions of science. Within a year of their discovery, there was great enthusiasm for the use of x-rays in medicine: one Chicago laboratory alone had already made some 1,400 images and there were hundreds of other locations around the world doing the same.

X-rays also created a public sensation. Advertised as the greatest scientific discovery of the age, an 1896 exhibition promised that the 'New Light' enabled you to see 'through a sheet of metal, a block of wood and even to count the coins within your purse'. The invasive power of x-rays excited curiosity and apprehension; a London firm advertised x-ray proof clothing for ladies, and a New Jersey bill sought to introduce a ban on the use of x-rays through binoculars. A poem from the 1896 *Electrical Review* hinted at their impact:

> I'm full of daze
> shock and amaze;
> For now a-days
> I hear they'll gaze
> Thro' cloak and gown – and even stays
> These naughty, naughty Roentgen Rays!

Such popular responses were not novel – they recalled the titillation of demonstrations of static electricity a century earlier – but the scale and expansion of interest was new. Manufacturers of x-ray apparatus proliferated to satisfy medical demand. As a growing number of operators were chronically disabled or even killed by excessive dosages from their tubes – some two dozen within the first year – government laboratories were drawn into measuring and standardizing exposures. Medical usage expanded even more rapidly during the First World War, where x-rays proved indispensable in diagnosing bullet wounds, broken bones, and heart and lung problems. The rays were useful beyond mere imaging, however. Proponents claimed that some 100 diseases – from birthmarks to syphilis – yielded favourably to x-ray treatment. So-called 'female problems', both physical and psychological, were thought to respond particularly well to radiation. X-ray treatments were used to treat depression and the menopause, and to remove facial and body hair. The machines were increasingly licensed for use in beauty parlours for hair removal, and in shoe shops to check the fit of shoes.

Just as x-rays could increasingly be controlled and applied, other radiations proved to have medical and commercial interest in the early twentieth century. In particular, radium (discovered by the Curies as an exceedingly minor component of granite) had interest for physicists, medical specialists and entrepreneurs. Radium, too, was touted as a therapy to treat problems such as asthma, arthritis, migraine, psoriasis and diabetes. Materials containing, or exposed to, radium were marketed in the forms of belts, water, creams, toothpastes, hair tonics and chocolate for invigoration and rejuvenation.

Unregulated commercial application of x-rays and radium introduced a growing health problem. Hundreds of women were injured or completely disabled by excessive exposure to depilation x-ray treatments before the Second World War. A significant fraction of factory workers painting watch dials with luminous paint containing radium developed cancers of the face, jaw or throat because they had licked their brushes. The chief medical examiner of New Jersey identified the cause, and the United States Radium Corporation subsequently closed in 1927 after legal action from its workers. Some 1,600 tons of ore and contaminated material remained on the site, which was remediated only at the end of the century.

Such dismal and incautious applications of scientific knowledge did not dominate government and public assessments. In fact, commercial employment of scientific knowledge expanded

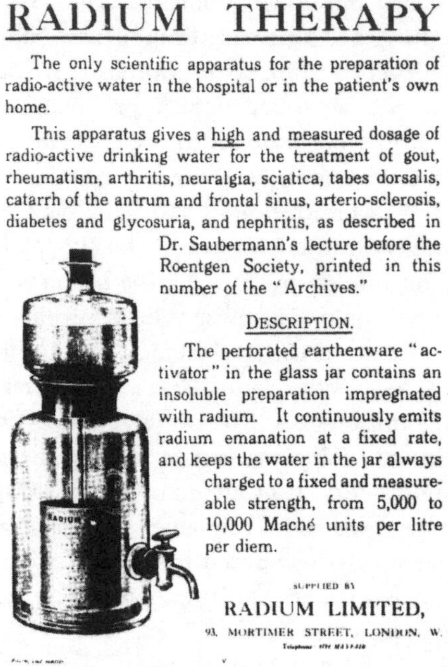

RADIUM THERAPY

The only scientific apparatus for the preparation of radio-active water in the hospital or in the patient's own home.

This apparatus gives a high and measured dosage of radio-active drinking water for the treatment of gout, rheumatism, arthritis, neuralgia, sciatica, tabes dorsalis, catarrh of the antrum and frontal sinus, arterio-sclerosis, diabetes and glycosuria, and nephritis, as described in Dr. Saubermann's lecture before the Roentgen Society, printed in this number of the "Archives."

DESCRIPTION.

The perforated earthenware "activator" in the glass jar contains an insoluble preparation impregnated with radium. It continuously emits radium emanation at a fixed rate, and keeps the water in the jar always charged to a fixed and measureable strength, from 5,000 to 10,000 Maché units per litre per diem.

SUPPLIED BY

RADIUM LIMITED,

93, MORTIMER STREET, LONDON, W.

Figure 8 Commercial radium therapy *c.*1930 (From Caufield, *c.*1989. *Multiple Exposures*. London, Secker & Warburg, Fig. 8)

dramatically from the late nineteenth through the twentieth century. As illustrated earlier, the linkage between knowledge, commerce and government grew slowly but steadily – so much so that the spirit of the age was the identification of progress provided through science. During the industrial revolution beginning in the eighteenth century, a relatively small handful of entrepreneurs had profited from the application of science to industry; during the nineteenth, governments increasingly monitored, controlled and supported such applications. By the twentieth century, however, industrial cultures adopted even more explicitly a scientific perspective. This can be demonstrated by three related threads: industrial research, scientifically informed commercial products, and public engagement with science through commercial rhetoric.

Each of these factors evolved rapidly during the first half of the century from earlier expressions. The nineteenth-century industrialization of physics through steam power, and of chemistry through chemical plants, had had the greatest economic impact, but newer science and technology rose in importance by the end of the century. From the 1830s steam railway networks began to expand; at mid-century, electric telegraphy permitted even more rapidly expanding networks to transform communication on a global scale. The British Empire, by the end of the century, was united by the telegraph and its associated engineering achievements: submarine cables linking Canada, Hong Kong and Australia. Other nations, identifying the same advantage from communication, rapidly built similar networks. From the late 1870s the invention of the telephone permitted communication by voice over increasingly longer distances. At about the same time electric lighting began to challenge gas lighting, and at the turn of the century a growing number of investigators became interested in communication by radio waves, a then still poorly explained phenomenon first demonstrated in 1888 by Heinrich Hertz (1857–1894) in Germany.

The new electrical industries in particular – telegraphy, telephone, lighting and radio – attracted an early and growing involvement by practising scientists. Some were simultaneously scientists, inventors and entrepreneurs. William Thomson (1824–1907) is an early example of the transformation of the role of science. One of the most influential physicists of his generation, Thomson was a long-time professor at the University of Glasgow, where he had transformed natural philosophy from a single-person department to one taught by a system of professors and lecturers, and supported by laboratory classes. His innovations were taken up at other institutions and became widespread in the teaching of science by the end of the nineteenth century. Thomson had an equal impact on industry, developing industrial connections throughout his career. He consulted for telegraphy firms from the 1850s. He invented, and commercially benefited from, apparatus for telegraphy, marine instruments and optical devices. Philosophical instrument maker James White (1824–1884) worked closely with Thomson to build his designs for electrical devices, compasses and sounding apparatus. In fact, when Thomson (now Lord Kelvin) gave up his university chair in 1899, he became Director of Kelvin & James White Ltd.

By the end of the nineteenth century newer electrical industries were rapidly expanding. The commercial products of the Westinghouse Company, founded as recently as 1886 and pioneering power via alternating current (AC), were able to fill an entire exhibition hall at the Chicago World's Fair in 1893. The innovation of AC power, which made possible efficient electric motors and the transmission of power from distant sources such as hydroelectric dams, was invented by Nikola Tesla (1856–1943). A Croatian with wide electrical engineering experience in Europe before emigrating to America in 1884, he worked for Thomas Edison (1847–1931), and later established independent laboratories and formed his own company before joining Westinghouse.

The notion of 'pure' and 'applied' science – a twentieth-century labelling – attempted to distinguish work that was increasingly intermingled. Just as the telegraphy industry had drafted Lord Kelvin to solve technical problems, the German Physikalisch-Technische Reichsanstalt brought together renowned academic physicists such as Hermann von Helmholtz with German engineers. Younger scientists such as Albert

Einstein were immersed in an environment that united abstract knowledge with industry. While contributing some of the most influential concepts to twentieth-century physics, Einstein (whose uncle owned a firm that manufactured electrical dynamos) began his career as a patent clerk, and patented his own inventions – including a refrigerator concept and automatic-exposure camera – as late as the 1930s.

Scientists and engineers such as William Thomson and Nikola Tesla were increasingly drafted as consultants or contract designers by the emerging (and rapidly merging) industrial firms in the final decades of the nineteenth century. By the beginning of the new century, however, Westinghouse and rival companies such as those directed by Edison and Werner Siemens employed a growing number of such scientifically trained researchers specifically to solve manufacturing problems and, even more novel, to seek improved designs. Innovation in the lighting industry, for example, required physicists to study power transmission systems, chemists to develop longer-lived lamp filaments, and biologists, psychologists and physicians to study the nature of vision. An increasing (but still small) number of commercial employees were scientifically trained. There was a clear gender divide, too: women were preferred for the subordinate technical roles of telephone operator and lamp inspector, just as they had been for a generation already for certain astronomical observations. Female workers were believed to be more patient, meticulous and accurate in such domains.

Science for promotion and modernity

And there was a further division of roles. Industrial scientists were identified early on as having a dual purpose: to improve competitiveness not merely through design improvements but also by supporting a new rhetoric of science. The promotional value of science suggested by the marketing of radium therapies was a broad transition made by early twentieth-century firms.

The promotional value of scientific industry had roots in literature as well as commerce itself. The early novels of Jules Verne (1828–1905) such as *From the Earth to the Moon* (1865) combined, for the first time, adventure with modern science. On the face of it, this was an incongruous juxtaposition for what had been portrayed as an objective, rational pursuit. While there had been earlier literature relating aspects of science to adventure, it had communicated a distinctly different tone. For example Mary Shelley's *Frankenstein* (1818), a cautionary tale revolving around the creation of life and the dangerous consequences of such power, imparts a sense of the sublime dimensions of science filled with awe, wonder and fear. In their own ways, accounts of a scientific adventure to Lapland for geodetic measurements, by Pierre Louis Maupertuis (1698–1759), and of South America by German naturalist Alexander von Humboldt (1769–1859), while also reflecting the sublime, largely limit science to collection, observation and recording.

By contrast, Verne added an active element to the mix, making technology (framed as applied science) as important as exotic environments to the plot. First appearing during the American Civil War, Verne's prolific output supported tales with enough scientific detail to astonish and convince. His genre was extended in popular 'pulp fiction' magazines such as *Argosy* (1880) and by other popular writers such as H. G. Wells (1866–1946) and Edgar Rice Burroughs (1875–1950). Burroughs, later famous for the *Tarzan* series of novels, wrote for pulp magazines and wrote successful serialized magazine stories and novels about adventures on Mars and Venus. Such locales had a scientific connection for contemporary popular readerships: Italian astronomer Giovanni Schiaparelli (1835–1910) from the late 1870s had reported observing *canali* (channels, mistranslated into English as canals) on Mars.

The public taste for scientific adventure expanded during the 1920s. Science was now commonly associated with adventure, exploration and excitement. American Hugo Gernsback (1884–1967), an entrepreneur in the early radio industry, introduced the first magazine, *Amazing Stories* (1926), specializing in science and fantasy. It was soon joined by other titles such as *Astounding*, *Future Fiction* and *Thrilling Wonder Stories*. His 'scientifiction', soon simplified to 'science fiction', proved a fertile magazine market, and was mirrored by contemporary cinema. *Buck Rogers*, *Flash Gordon* and other serial films broadened audiences between the world wars.

Such magazines provided a link between not only science and adventure, but invention and industry, too. The incongruity introduced by Verne seventy years earlier no longer seemed remarkable. Consider, for example, a 1939 issue of *Startling Stories* that combined science adventure fiction with 'The Life Story of Robert Millikan', an American physicist (1868–1953) best known for having determined the charge of the electron: such careful experimental toil now equated to popular adventure.

This genre of fiction forecast a scientific and technological future closely aligned with society. Men's magazines such as *Popular Science* illustrated a growing fashion in Western countries for a hobbyist interest in science and invention. Rather than adopting a tone of intellectual improvement and social betterment like their nineteenth-century counterparts, they combined articles on scientific discovery with home workshops and car repair. Science was increasingly identified as a clever skill, as useful as knowing how to tile a floor, build a radio or clean a carburettor.

Such popular magazines also adopted and disseminated rhetoric linking science with a positive social future. *Everyday Mechanics*, for instance, was relaunched as *Everyday SCIENCE and Mechanics* and adopted a masthead that reflected its theme. It depicted radio masts, electrical machines and light bulbs; airships and airplanes; a gleaming city with skyscrapers in front of a rising sun; and even factories and smokestacks – a positive sign of active industry that belied the ongoing economic depression. Science by the 1930s, in fact, was increasingly seen as an integral part of society – a force that was forecast to improve society dramatically, transforming it from the dirty industrial landscape of the nineteenth century to a gleaming, clean and efficient world of the twentieth century. For the first time, science equalled modernity.

Public understandings of science were arguably shaped as much by popular accounts as by formal exposure in schools. But such idealistic understandings were mirrored, (as we have seen in the case of radium therapies) by commercial rhetoric. Perhaps the most pervasive form of scientific fiction is in the form of advertising. Commercial use of science in advertising paralleled the rise of scientific periodicals and fiction. First, science increasingly was represented as a source of both discovery and invention. Companies that could promote themselves as 'scientific' attained an aura of authority and modernity.

Consider pharmaceutical products, for example. While large firms (e.g. Proctor & Gamble, founded in the USA in 1837, and Bayer AG, founded in Germany in 1863) were creating research laboratories at the end of the nineteenth century, the marketing of medications only later became linked with scientific authority. Indeed, the nineteenth-century norm suggested older alchemical or religious trusts: secret recipes, family expertise or revelation (as for the 'Wonderful Dream Brand Salve Company').

By the turn of the century medical authority was increasingly invoked in order to market medications. 'Dr J. D. Kellogg's Asthma remedy', one of a spectrum of products supported by private spa and exercise regimen, is typical. After the First World War, though, advertising increasingly took up scientific connections, whether or not the firm undertook scientific research and development. Print advertisements between the wars were increasingly imbued with scientific accoutrements. 'Discovery', 'advance' and 'innovation' were coupled with scientific terms and statistics. Their illustrations depicted (white-coated) scientists and apparatus. Indeed scientists, having a long status as objective fact-finders, were more frequently enrolled to endorse products. Thus a 1941 *Chesterfield* cigarette advertisement depicted researchers measuring nicotine and quoted statistics for the brand's relatively low nicotine level and, a decade later, the 'scientific facts in support of smoking'.

This marketing slant was equally popular in the physical sciences. The ubiquity and mundane nature of such claims by the middle of the century is illustrated by the printing company Sperry (Figure 9), which vaunted '200 research people with over

Figure 9 Sperry Research advertisement, 1947 (Sperry Research)

2,000 years of experience' against a background of oscilloscope traces, microscope, chemical apparatus and time apparatus.

The early twentieth century, then, brought a new proximity between science, government and commerce with mixed outcomes. Expanded scientific understandings of physics, and new medical powers to diagnose and treat, were paralleled by popular enthusiasms and commercial exploitation. Science became a more obvious part of popular culture between the wars, and its special cachet during the century was correlated with the world wars. The disparate social consequences of scientific knowledge were also foregrounded by warfare in the twentieth century.

Chemists' war

The ability to wage war, generally intended to protect or extend a nation's resources, power or way of life, has been a tool of most societies at least during periods of conflict. During the twentieth century, however, science was increasingly applied to all aspects of warfare. By late century, the superpowers – a term defined by the new scientific and technological expertise acquired with the Second World War – invested growing portions of their economy into weapons development.

At the beginning of the century, the scientific contribution to these developments was limited in scale and range. For instance the design of optical instruments was largely a continuation of an established culture of instrument making. Newer binoculars, rangefinders and signalling equipment, although sometimes designed by scientists, were still fabricated by artisans just as eighteenth-century sextants and surveying transits had been. New explosives developed during the 1880s, on the other hand, were manufactured on an increasing scale. Alfred Nobel (1833–1896), the inventor of dynamite and gelignite, initially argued that his explosives would end warfare. Seeking a more certain legacy, however, he willed most of his estate to establish the Nobel Prizes for physical science, chemistry, medicine, literature and international fraternity.

Awarded in 1901, the first Nobel Prizes coincided with the increased application of technology to support the Boer War (1899–1902), Russo-Japanese War (1904–1905) and First World War (1914–1918). Mass production of new technologies – tanks, aircraft, submarines, radio communications and improved artillery – collectively transformed battle. Beyond mere engineering and scale, however, the First World War was a turning point in marshalling scientists as key participants.

On a grand scale, the war was reliant on chemistry. The supply of chemical munitions was central to the waging of the war and the creation of immense munitions factories combined science and industry with government management. The Gretna Explosives Factory, for example, covering a 6,000 acre, nine-mile swathe across the Scots-English border, was the largest such munitions works in the world up to that time. The largest of eighteen sites operated by Britain's new Ministry of Munitions, the Gretna factory employed over 16,000 workers to manufacture cordite (the latest smokeless explosive), and was managed by chemists from around the British Empire. Many of them, products of the expansion of universities over the previous generation, were withdrawn from military service or seconded from academic posts and drafted off to the stream of new factories. Government-directed science also was fostered in the new Department of Scientific and Industrial Research (DSIR, 1917).

Gretna represented a new style of waging war: one that relied on scientific expertise. The 'chemists' war' eventually mobilized well over a thousand technical workers with chemical backgrounds – a considerable fraction of Britain's scientific labour. Such large-scale factory management was a new and widely noted government activity. Gretna required not merely factory facilities and scientific expertise to design chemical processes, but a rational and quantitative approach to management, too. Female workers exposed to cordite by skin or ingestion developed yellow skin and reddish hair, and were known as 'canary girls'; the detection and avoidance of such industrial illnesses led to new legislation for occupational health and safety. Government ministries also planned, created and managed a neighbouring town. Efficient production, occupational health and civic planning broadened the chemists' remit to wider social planning. Such wartime examples encouraged the growth of so-called 'scientific management' and 'efficiency engineering' in Western countries during the 1920s, and an injection of a scientific approach to government administration.

Of equal consequence for post-war science was the development and battlefield deployment of poison gas. The French first used a tear gas in fragmentation shells in the first months of the war. The Germans employed a similar chemical, also with little military effect, until the spring of 1915. However, with the assistance of Fritz Haber (1868–1934) of the Kaiser Wilhelm Institute for Chemistry, the German military developed methods of deploying chlorine gas, a lethal by-product of dye manufacture by German chemical companies. The British and allies developed battlefield protection – initially moistened cotton – and began to retaliate with chlorine that autumn. Most of the wartime deaths and disablement from poison gas were attributable to phosgene, first used by the French at the end of 1915. However Mustard gas, introduced by the Germans in 1917, became more widely known because it was more likely to incapacitate than to kill outright. At the end of the war, the United States and Britain had large unfinished plants for new poison gases under construction. Seven countries, dominated by Germany, France and Britain, had released some 150,000 tonnes of poison gas, leading to an estimated 90,000 deaths and 1.2 million casualties.

Invalided soldiers, blinded or disabled by compromised lungs, became one of the most visible outcomes of the war for the public. International abhorrence led to the Geneva Protocol (1925) banning the use of such chemical weapons, as well as bacteriological weapons such as anthrax. Even so, stockpiling and further development was permitted by the treaty, and several of the original combatant nations and others continued to do so until the 1970s. For example, the USA, Russia, Britain and Germany accumulated – but did not use – stockpiles of poison gas during the Second World War (including Germany's more lethal nerve agents).

By the 1930s, the public backlash to the 'chemists' war' had at least two components: first, horror at the use and effects of poison gas and, second, a growing conviction that chemical firms had profiteered from the war – effectively uniting scientific knowledge inappropriately to gain an industrial and economic windfall. The rise of the phrase 'merchants of death', first applied to Alfred Nobel, reflected an inter-war popular distaste for militarily oriented industry and research.

Even so, science was seen equally as a power for national benefit. Scientific support for the war had been strong from each nation. Albert Einstein bucked the trend by evading contribution to the German war effort, unlike most of his contemporaries. Enthusiasm in Britain was initially so strong that war work at the national laboratories, for example, was seriously hampered as technical staff enlisted to fight in the trenches. The strong national allegiances were in marked contrast to the position of scientists over a century earlier, when European scientists had been relatively free to communicate and even visit each other during the Napoleonic Wars. The outcome of the First World War was the recasting of internationalism in science. While scientists became more nationally oriented, they also adopted a more visible professional and social role in their countries. Germany, for example, was excluded from participation in international scientific conferences for a decade. Scientifically isolated, a fraction of its scientists argued that science itself had been tainted by social factors. Physicists Phillip Lenard (1862–1947) and Johannes Stark (1874–1957), for example, argued that good 'Deutsche Physik' ('German physics') must oppose theoretical so-called 'Jewish physics', epitomized by Albert Einstein and his 'scientific fraud' of relativity. Such scientific claims, categorizing science along national and cultural lines, supported the rhetoric of Adolph Hitler when he came to power in 1933.

Physicists' war

If the First World War has been dubbed 'the chemists' war', then the Second World War is justifiably called 'the physicists' war'. Unlike the lead-up to the First World War, when governments had found supplies such as optics, explosives and even dyes in short supply, planning for another world war began following Hitler's rise to power. Scientific development was now even more overtly supported. Radar (an acronym for 'RAdio Detection And Ranging', coined in 1941) was developed from the mid-1930s in Britain, Germany, France and the USA. The technology, intimately reliant on physical theory and experimentation, developed rapidly through the war and beyond.

Of equal consequence during the war and afterwards, though, was atomic energy. The phenomena of radiation and atomic properties, as discussed on page 113 in the early observations of 'alphabet rays', were a subject of enthusiastic research through the century. Research by physicists and chemists in widely separated laboratories culminated, in 1939, with the first evidence for the artificial disintegration of an atomic nucleus along with a corresponding release of energy. The experiment relied on the *neutron*, a subatomic particle identified just six years earlier. Radioactive elements, it was determined, decomposed with the release of such particles and transmuted into another element of lower atomic weight. Some, such as the heavy uranium nucleus, could be divided by collision with an energetic particle.

The engineering interest in this arcane process relied on the fragments that were released: if more than one neutron were released on average with each nuclear division, then those neutrons could cause further divisions, and potentially a cascading and increasing number. This nuclear *fission* was understood by physicists around the world to promise a useful source of energy if this *chain-reaction*, or cascading sequence of fissions, could be engineered. At the beginning of the war, then, atomic research followed discrete and secret national routes in those countries that had the resources to pursue the research.

Britain, principally through the group under Ernest Rutherford at Cambridge, had hosted such research for a half-century. Refugee physicists from Germany bolstered their expertise and, familiar with ongoing research in their country, provided a sense of urgency. The first careful study of the potential of atomic energy was made through the code-named MAUD committee appointed by government in 1940. The small collection of physicists, supported by chemists of Imperial Chemical Industries (ICI) concluded that the development of an atomic bomb based on uranium was very likely feasible within two or three years, and that longer-term development of atomic energy as a source of power appeared promising. Given the threat of German invasion and the limited manpower available, though, they recommended moving the project to either the United States or Canada.

In America, Albert Einstein had authored a letter to President F. D. Roosevelt in August 1939, but research activities had been relatively uncoordinated. The government decision to fund an American bomb project was guided by two highly placed administrators, who subsequently played an important role in guiding the field. Vannevar Bush (1890–1974), an engineer who had worked on submarine detection during the First World War, was Chair of the National Defense Research Committee (NDRC). When he became director of a more powerful coordinating body, the Office of Scientific Research and Development (OSRD), created in 1941, he appointed James B. Conant (1893–1978), President of Harvard, as scientist responsible for overseeing explosives development. When they learned of the MAUD reports they convinced Roosevelt to fund the most expensive scientific project of war, known afterwards as the Manhattan Project. Collaboration between the United States and Britain was diffident and sporadic. The British felt the American efforts lagged at first, while the Americans mistrusted the international and industrial complement of workers sponsored by Britain. Hesitant and periodically excluded, the British correspondingly focused their efforts on collaboration with Canadian scientists.

The atomic bomb project was unprecedented in its scale and employment of scientific expertise. To maximize the likelihood of success in a useful time period, several scientific and technological threads were pursued in parallel. The most straightforward but time-consuming method was to purify one radioactive isotope of uranium to create a bomb. This was undertaken by three methods, each requiring vastly expensive large-scale plants built at Oak Ridge, Tennessee. The atoms of slightly different mass could be separated by a diffusion process, electromagnetic separation or centrifuge. Pursuing a fundamentally different approach, at the University of Chicago a group under Italian physicist Enrico Fermi (1901–1954) experimented to design the first *chain-reactor*, or *pile*. When sufficiently scaled up, such a lattice-work of uranium and graphite could transmute uranium into a never-before observed element, plutonium. Plutonium, it was hoped, could also yield a bomb, but would require a more complex triggering design. The designers of the uranium and plutonium bombs were led by physicist Robert Oppenheimer (1904–1967) at Los Alamos, New Mexico. The entire American, Canadian and British development effort was directed by the American military under General Leslie Groves.

Each of these separate developments required fundamental scientific research. For the first time, the military leaders were required to manage the creation of novel forms of factory – involving new phenomena such as high-intensity radiation and its little-known effects – in collaboration with fundamental scientists and industrial engineers. The Du Pont chemical company built and operated the large scale Hanford reactors while the scientists were still investigating the fundamental physics that underlay them. At other sites Westinghouse, General Electric, Kellogg and other companies managed production and the scientists and other technical workers.

By early 1945 an adequate quantity of uranium had been separated and sufficient plutonium had been manufactured in the production reactors for a few bombs. It was by then apparent that the German war would soon end, and so the original impetus – to preclude the use of an atomic bomb by Germany on the allies – had disappeared. The project scientists largely favoured a delayed use, a warning demonstration, or no use of the bomb at all. They were overruled by the military, who argued that the use of one or more bombs on Japan could quickly end a war that would otherwise require another year and perhaps a million American soldiers' lives. That August, one uranium bomb (on Hiroshima) and one plutonium bomb (on Nagasaki) were dropped, with an estimated loss of some 150,000 Japanese lives by the end of the year.

For both scientists and the public, the Manhattan Project highlighted a new ethical dimension for science, a theme picked up below.

Political dimensions

The wider public and their governments learned other lessons. The development of the atomic bomb consolidated international power for America. Britain, nearly bankrupted by the war, sought to maintain its status in the post-war world and undertook a programme to develop its own nuclear weapons. The Soviet Union, even further weakened and more immediately concerned with invasion, did the same. In France, the scientists who had been part of the Anglo-Canadian project similarly developed their country's nuclear capabilities. For each of these countries, the bomb came to represent national prestige and power, with direct economic and political consequences.

For a time, the atomic bomb represented something more, especially in the USA: exhilaration at the conclusive end to the war and at the demonstration of technical supremacy. The cultural echoes are suggested by post-war music ('Atomic Boogie', 'Atomic Cocktail' and even 'Jesus Hits Like an Atom Bomb'), chewing gum ('Atomic Fireball, with Red Hot Flavor') and even swimwear. (The 'bikini', introduced in 1946, got its name from the atomic tests on Bikini Island in the South Pacific; its French designer reasoned that it was hot, tiny like the atom and likely to produce a burst of excitement like the bomb.)

More enduringly and surprisingly, perhaps, the atomic bomb provided a new positive gloss for science and scientists. Public representations of science through the 1950s portrayed it as an all-powerful solution to problems. As a result, scientific advertisements gained a new potency for industry. The solutions it seemed capable of providing also extended beyond science and technology to social improvement. Scientists, no longer vaunted as isolated and objective seekers of truth, were now the providers of national security and improvement, increasingly integrated into business, commerce and government as advisors or creators. The terms 'rocket scientist' and 'atomic scientist' became a short-hand for all-powerful authorities.

The high secrecy and administrative organization of the Manhattan Project also provided a model for post-war government-managed military research. New national laboratories took over where their wartime equivalents had left off. In America, government-funded atomic sites such as Los Alamos, New Mexico, Oak Ridge and Hanford were managed by industrial contractors. Institutions such as the University of California at Berkeley increased their military contract work and became reliant on government funding. A growing wave of scientific research contracts were sponsored by the armed forces for missile systems, battlefield surveillance, communications research and other technologies potentially of military value.

Not all scientists welcomed the new scale, security concerns and subjects of post-war work, and emigrated to other fields or locations. A post-war song by physicist Arthur Roberts lamented the new conditions for sub-atomic research:

> Take away your billion dollars, take away your tainted gold,
> You can keep your damn ten billion volts, my soul will not be sold,
> Take away your army generals, their kiss is death, I'm sure.
> Every thing I build is mine; every volt I make is pure.
> Take away your integration; let us learn and let us teach,
> Oh, beware this epidemic Berkeleyitis, I beseech.
> Oh, dammit! Engineering isn't physics, is that plain?
> Take, oh take, your billion dollars, let's be physicists again.

Other scientists shifted disciplines; the field of 'health physics' combined the physicists' approach to measurement with the biologists' concern for understanding how living systems cope with radiation. This new terrain and new form of government-funded science extended beyond military applications, too. The Manhattan Project's deferred promise during the war had been atomic energy for civilian benefit. When the war ended, many of its scientists were eager to contribute to this more positive form of social outcome.

The development of civilian nuclear power in America, Britain and the Soviet Union during the 1950s was nevertheless driven by various government interests rather than industry. Atomic energy research remained a highly classified field for a decade after the war oriented towards improved nuclear weapons and gradually exploring the potential for nuclear power generation. Echoing the prohibitions after the First World War, Germany was prevented from engaging in experimental research in the field until 1955. In America and Britain, the administration of radioactive materials and technology became government monopolies under the new Atomic Energy Commission and Atomic Energy Authority, respectively. In France, the Commissariat à l'Énergie Atomique (CEA) adopted a similar role. Incidentally, the term *atomic* was gradually superseded in popular usage by its more accurate counterpart, *nuclear*, by the 1960s. For a time, *atomic* evoked weapons in public discourse, while *nuclear* was free of such associations; this linkage was later reversed.

Nuclear energy represented a new style of engagement between governments and science. The field was wholly funded and directed by government organizations. Formation of the academic discipline, held back by secrecy during the early phases of the Cold War, eventually began in the late 1950s, again with direct government support in the USA. Commercial influence in the field remained restrained into the 1960s, in part because of the engineering uncertainties associated with the technologies.

Apart from weapons, nuclear technologies appeared unpromising to many scientists immediately after the war. Uranium stocks appeared too small to support power generation for more than a few decades, so a variety of reactor designs were studied. All create heat by a controlled chain reaction. This heat is extracted and transferred to drive relatively conventional turbine systems and so yield electricity. For each reactor concept, key physical properties – such as the effects of radiation on the materials – were at first inadequately understood. The economics were consequently uncertain but hoped to be competitive with other sources of energy in certain regions. In Britain, expected to face shortages of coal miners, nuclear energy appeared to offer a relatively short-term solution. In all countries, however, nuclear energy was developed as much for its political benefits. The earliest civil power station, Calder Hall (1956) in England, was optimized for plutonium production rather than electrical power generation. Shippingport (1957), its American equivalent, was a land-based version of the reactor designed for the first nuclear naval vessels. But nuclear science augmented national and commercial status, both important to post-war administrations.

Indeed nuclear energy was an important demonstration of more than scientific expertise. It was cited to support claims of the superiority of the respective economic systems of America and the USSR. For the Americans, the Manhattan Project was the model for not just a post-war nuclear programme, but an enduring range of government-funded programmes for defence, health and research. Reflecting termination of the Second World War by powerful science, new scientifically informed government programmes adopted the war metaphor: the War on Poverty (Lyndon Johnson, 1964); the War on Cancer (Richard Nixon, 1971); the War on Drugs (Ronald Reagan, 1982). Such initiatives extended the techniques of physical science and engineering to the social and biological sciences. Less aggressively, Vannevar Bush, as one of the key administrators of the Manhattan Project, could argue that 'the endless frontier' of science could provide national prosperity to a free country. He called for basic research by unconstrained scientists with government funding to continue the rapid technological development achieved during the war in peacetime. One result was the National Science Foundation (NSF, 1950). His wartime colleague and advisor to the NSF, James Conant, extended this vision to the discipline of history of science. He argued that science was an intellectual pursuit that could not be influenced for long by social or economic forces. The Harvard Case Studies in Experimental Science instituted by Conant promoted this philosophical emphasis for American history of science.

At the same time, Conant's perspective suggested that science was intrinsically effective in problem-solving, and that funded research would inevitably lead to successes (Figure 10).

The Manhattan Project represented a seductive example to such administrators, and motivated post-war 'big science' projects. More cautiously, subsequent historians would counter that the definition of success can be difficult to agree upon. Technical accomplishment, for example, measured according to efficiency or destructive capacity, may have no simple link with social benefit or popular support.

Ironically, too, the policy of post-war alliance between American government and science had considerable similarity to

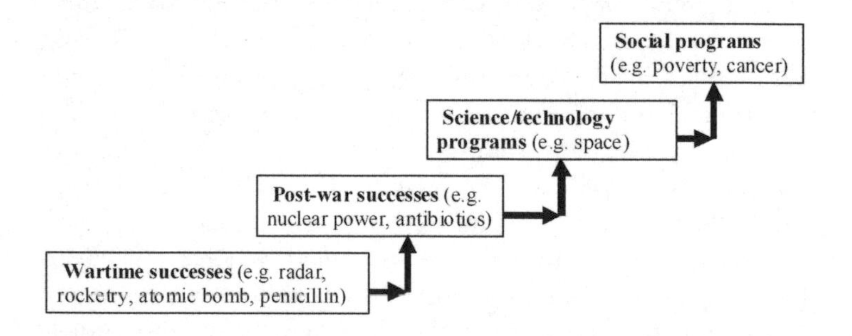

Figure 10 Late twentieth-century trends in American science suggested by James Conant's perspective (S. Johnston)

the model practised in the USSR. For many Soviet commentators, the political philosophy of Marxism represented a science. They argued that it was based on a detailed socio-economic theory and led to testable predictions of historical, social and political change. Indeed, it provided a philosophy of scientific investigation, 'dialectical materialism'. According to this understanding, the Soviet implementation illustrated a planned and rational economy. In particular, scientists would be centrally organized to work efficiently on problems of national significance. Intellectual problems would be selected rationally and research, without wasteful commercial duplication and profit-taking, would be efficient and oriented towards the common good. Science, they argued, would be made both intellectually and socially progressive by design. And unlike Conant's view of the independent and dispassionate scientist, the Soviet vision of the history of science saw it as being shaped by collective research and consensus, and directed by economic and social forces.

For both countries, science and technology represented a new battleground during the Cold War. The production of ever more powerful and numerous nuclear weapons became a visible indicator of relative progress. The 'space race', begun with the launch of the Soviet *Sputnik* (1957) and ending with the American lunar excursion module *Eagle* (1969), was the public face of competition founded on the development of intercontinental ballistic missile systems to carry nuclear warheads.

Ethical dilemmas

In parallel with the rise of politicized science came new ethical concerns. These, too, came to a head during the Second World War. After the war, the work of individual scientists and the directions taken by science policy were both influenced by this new framework.

The question of the ethical conduct of scientists had been expressed at least occasionally from the beginning of the century. As suggested by Figure 11, the application of science to warfare was particularly criticized; for example, the work of Fritz Haber and other scientists on poison gas was debated, particularly outside Germany, after the First World War (although Haber controversially was awarded the Nobel Prize in Chemistry in 1918 for his work with Carl Bosch on nitrogen fixation

Figure 11 'Modern science and prehistoric savagery' (Wil Dyson, *Sydney Morning Herald*, 1912)

processes, of immense value to international agriculture). Organizations dedicated to scientists as professional workers and as members of society had, in fact, begun after that war: the most visible example was the Association of Scientific Workers in Britain. But the scientific capabilities developed during the Second World War suggested a more intense and direct culpability. The atomic bomb was a weapon even more indiscriminate than poison gas: it would destroy not just military installations and personnel, but civilians, too.

The previous two sections sketched some details of the atomic bomb project. A further layer of decision-making during the Manhattan Project concerned its appropriate use. The organization of the project encouraged just this kind of segregation of activities and responsibility. The initial British and American development had begun with the assumption that Germany's scientists were already engaged on such a project. In this them-or-us context, nearly all the scientists worked enthusiastically on the project. Security and military responsibility for project management and decision-making encouraged this division of labour. And, for many of the scientists, there was a positive incentive: the novelty and importance of the scientific questions raised by the bomb.

By 1944, however, it was becoming clear that the war in Germany would end before the bomb was ready. Members of Project Alsos, an offshoot of the Manhattan Project that sought to discover German progress in developing an atomic bomb, advanced into Europe following the D-Day invasion. They later concluded that the German effort had been ineffective, although the scientists under Werner Heisenberg had continued toward their goal until the end of the war.

Some of the Manhattan Project scientists questioned the need to complete, and deploy, such a weapon when the primarily military threat had been removed. Japan, still waging war, represented no threat of invasion to America, at least, and was securely constrained by allied forces. They argued that one option would be to demonstrate the bomb to international observers, including the Japanese ambassador.

The military decision to use the atomic bombs on Hiroshima and Nagasaki was based on utilitarian arguments. It was estimated that defeating Japan using conventional forces – involving either a naval and air blockade, or via island-by-island invasion – would create large numbers of casualties on both sides. The American experiences of invading Guam and the Philippines had suggested that Japanese forces would not surrender, and that American casualties could range from one-quarter million to one million soldiers. In the context of wartime propaganda that depicted the Japanese as an indefatigable and brutal enemy, their value in the utilitarian arithmetic was further reduced.

Utilitarianism: the ethical tradition that judges acts by their utility in producing the greatest good for the greatest number.

There were wartime precedents, too, for this scale of military response. The Allied fire-bombings of Dresden and of Tokyo, both in February 1945, had caused tens of thousands of deaths, principally of civilians. According to such sums, the deaths of an estimated 100,000 in Hiroshima and 60,000 in Nagasaki in the months after the bombs were dropped that August could be rationalized or defended. (Incidentally, these figures represent half of the estimated total deaths that eventually resulted from injuries and radiation illnesses.)

The morality of scientific warfare was the most public ethical question following the war, but others were of equal concern to practising scientists. Immediately following the use and announcement of the bombs, a censorship policy was introduced to restrict further release of information in the USA,

Britain and Canada. The American McMahon Act of 1946 extended this embargo on information to its former allies, too. The secrecy of scientific information became an ethical issue. The post-war identification of science with political ideologies, as discussed in the previous section, motivated patriotic scientists to preserve their country's advantage by withholding knowledge that could be useful for military purposes.

In such an atmosphere, the ethics of scientific behaviour were ripe for debate. While for some post-war scientists nuclear research allowed them to act 'like children in a toy factory' (to quote Alvin Weinberg), others felt burdened by new responsibilities. Physicist Richard Feynman, for example, who had led a theoretical group at Los Alamos during the war, recalled being devastated by the realization of the destructive power of the bomb, and the futility of planning a life after its use.

The wartime groupings of scientists had banded together to form, for example, the Atomic Scientists of Chicago and the Atomic Scientists' Association in Britain. A visible expression of the social conscience of such scientists was the *Bulletin of the Atomic Scientists*, in the USA, and *Atomic Scientists' News*, in the UK, both launched after the war. These post-war organizations sought to educate and influence their governments to develop atomic energy for peaceful rather than military purposes and initially favoured international control of atomic weapons and dissemination of their knowledge to all countries. Collectively they supported the free release of atomic information to prevent a monopoly of knowledge and danger of imminent war – a policy strongly promoted by Niels Bohr in the post-war years. Even so, the scientists' aims, and Bohr's advocacy, were widely judged as naïve by politicians and policy-makers of the period.

The ethical debate concerning nuclear warfare was revitalized by development of the so-called hydrogen bomb, based on nuclear fusion, from the mid-1950s. Its testing by America, the

Soviet Union and Britain was accompanied by concerns about radiation exposure of civilians. Contention also focused on the moral justification for this much larger, and so even more indiscriminate weapon, and on the arms race that it fostered. Here again, the ethical issue was raised of national allegiances for scientists. The 1958 founding of the Campaign for Nuclear Disarmament (CND) in Britain brought these issues to the public. CND was the first expression of public opposition to scientific research and application.

Scientists' sense of responsibility for such developments varied. Robert Oppenheimer, Director of the Los Alamos team developing the first atomic bombs, famously observed, 'It is not the scientist's responsibility to determine whether a hydrogen bomb should be used'. In the same vein, Lewis Wolpert, emeritus Professor of Biology as Applied to Medicine at Imperial College London, wrote, 'Science is value-free, as it explains the world as it is. Ethical issues arise only when science is applied to technology – from medicine to industry'.

This neat division between 'pure' and 'applied' science is seen as rather too convenient by some critics. Indeed, the separation of science into these two parts was an invention of the early twentieth century and accompanied the creation of commercial scientific laboratories. Such labs were often presented as a natural step in the discovery and application of knowledge in the tradition of Francis Bacon. According to this understanding, then, the ethics of an application could be isolated to decision-making somewhere along this chain from theoretical conception to end result. There is no doubt that ethical qualms were submerged or divorced from scientific research for many of the participating scientists during the three years of the Manhattan Project, but ethical quandaries were revived as the availability of the weapons loomed. In the post-war world, the ethical dimensions shifted. Subsequent nuclear weapons proved ever-more destructive but arguably had less emotive impact; on the other hand, critics increasingly cited the illogic of military action on this vast and internationally threatening scale.

While the atomic bomb first focused wide attention on the ethical issues in science, other concerns took its place over the following decades. Many moral questions had, at their core, the sense of scientists overstepping their boundaries and 'playing god'. An example from medicine, greatly empowered by new medicines and technologies after the Second World War, was the heart transplant operation. First attempted in 1967, the procedure led to the death of its patients within days or months until the 1980s, when the problem of tissue rejection was largely overcome. With the advent of organ transplantation and improvements in the technology of life-support, medical practitioners gained dramatically in the power to extend or preserve life. Ethics committees, sometimes referring to themselves self-consciously as 'god committees' during the 1960s, were established to made decisions about the use, or termination, of such technologies for particular patients.

A more recent and even more contentious field prompting questions about the moral direction of science is genetic engineering. From the late 1990s, in particular, genetic research seeking to modify plant and animal species has been popularly associated with moral tales such as Mary Shelley's *Frankenstein* The issues in these cases are arguably even broader, ranging from utilitarian concerns (e.g. is it worth risking unsuspected outcomes in return for improved efficiency in food production?) to the moral limits of scientific research (e.g. is it right to mix genetic material from different species?). While recent and ongoing, such current subjects are also valid and fruitful topics for the historian of contemporary science.

Both applications of medical ethics – in judging the appropriateness of research and of treatments – had been fostered by the experiences of the Second World War. The medical experimentation of researchers such as Josef Mengele in the concentration camps of Nazi Germany, and of Japan's Unit 731, using prisoners of war as subjects for research in biological warfare, led to the refinement of post-war codes of ethics for medical practitioners and researchers.

Despite the need to curb such flagrant excesses of misapplied science, there has been considerable debate about ethics and authority. Should doctors be seen as the competent, and perhaps sole, authority in decisions of life and death?, it is asked. And should not researchers have the unchallenged right to gather information unconstrained by society, in order that they can eventually deduce important connections and new discoveries? Ethics committees increasingly have challenged such traditions. They argue, for example, that research data can be misused if not appropriately constrained to a defined objective. On the other hand, the researchers themselves chafe at this increasing regulation by the non-expert.

Ethical sensibilities about professional conduct, both inside and beyond science, have been heightened in recent decades. The examples mentioned here barely touch upon the issues affecting good practice for scientists (involving their research aims, their interactions with the subjects and colleagues with whom they come in contact, the confidentiality and dissemination of information) and also for historians of science.

Technological traumas

Science and its practitioners had periodically aroused public concern before the twentieth century. Isolated examples include the 1791 burning of the home of Joseph Priestley by a mob (some of whom linked his science with his theology and radical politics), popular ridicule of Darwin's theory and the Victorian anti-vivisectionist movement. Following the Manhattan Project, however, the ethics and achievements of science gained wider-scale and more concerted disapproval.

Criticism of the original atomic bomb was muted but its dramatically more powerful successor, the hydrogen bomb, focused increasing public concern. An illustration is the formation of the Campaign for Nuclear Disarmament (CND, 1958). A year earlier the first widely reported nuclear accident had occurred. Largely because of physical properties of irradiated graphite that were originally unknown to its designers, the Windscale plutonium production reactor in England caught fire. The release and fallout of radioactive smoke destroyed local milk supplies and gained international attention. Less known but significant nuclear spillage, leakage, contamination, overheating and explosive incidents occurred in Ontario, Canada (1952), Mayak, USSR (1957), California (1959), Idaho (1961), Michigan (1966) and Scotland (1967). Although some led to fatalities, there was relatively little public concern in a cultural atmosphere largely supportive of the new technology.

By the 1970s, however, criticism of civilian nuclear power was rising. The public had gradually come to associate nuclear weapons and the American and Soviet policies of mutually assured destruction with civilian nuclear power. The expansion of commercial power plants increasingly affected nearby local communities in decision making, and government regulation and support of the industry was challenged in several countries. The 1979 partial melt-down at the Three Mile Island nuclear plant in Pennsylvania, coupled with the coincident release of the film *The China Syndrome*, consequently provoked condemnation of the industry and the technology itself. The catastrophic melt-down and explosion at Chernobyl, Ukraine (1986) consolidated such concerns. These disastrous failures ended what nuclear engineer Alvin Weinberg (1915–2006) dubbed 'the first nuclear era' in most countries, halting the commissioning of many planned nuclear plants. (By contrast, nuclear energy dominated electricity generation in France by the end of the century).

The backlash against nuclear energy – vaunted as a key to social prosperity during the early 1950s but increasingly reinterpreted as an economic failure and public health concern by the 1980s – characterized a wider public re-evaluation of science and scientists. Just as the post-war status of scientists had been carried aloft by their wartime achievements, their esteem plummeted along with some of the products attributed to them. Ironically, in the context of global climate change, nuclear power may be rehabilitated along with the scientists who have contributed to it.

This collective tarring of reputation would not have been notable if isolated to a single scientific domain, but through the late twentieth century similar popular reassessments were made about other fields. As with nuclear energy, attention focused on technological failures that had been linked to novel science. It is worth noting that this close affiliation between science and technology was a creation of the century. As we have seen, the rise of commercial and government linkages with science highlighted the application of scientific knowledge. This in turn encouraged understandings of 'pure' and 'applied' science and the conviction, expressed by Vannevar Bush, that unconstrained research led naturally to beneficial technological products. But technology had not always been seen as subordinate to science; inventors had always employed multiple sources for creative ideas.

Simple definitions – increasingly seen as unrealistically simplistic by scholars – suggest that science concerns investigation to gain understanding of the natural world, and that technology concerns tool-making and invention. From the cases already examined in this book, however, it is apparent that episodes of scientific achievement have generally entailed physical manipulation and experiment, which require technical skills and a motivating purpose. Similarly, the adaptation of the physical world to human needs generally benefits from deep understandings. Science and technology are related, but both have often been influenced or shaped by philosophical orientation, commercial pressures, religion, innovation and social interaction.

Nevertheless, the cultural role of science increasingly was determined by technological successes and failures during the late twentieth century. The ability of science to monitor and diagnose problems, as well as public reaction to those problems, were considerably augmented after the war. For example, the Great London Smog of December 1952, a five-day period of stagnant pollution, was apparent to Londoners as a much heavier than usual fog. Similar events had occurred in Belgium (1930) and Pennsylvania (1948), and became endemic in car-centred cities, notably Los Angeles. The adverse medical effects were nevertheless detected by British health officials who estimated some 4,000 excess deaths from pneumonia, bronchitis, tuberculosis and heart failure, and an estimated 8,000 excess deaths afterward. The revelation, and public outcry, led to a Clean Air Act four years later which banned dirty fuels in industry and urban centres.

Acid rain, attributed during the 1970s to concentrated chemical industries in regions such as the north-eastern USA, proved even more difficult to identify and link to human activities. During the following decade, depletion of the ozone layer was detected by satellite observations; it was ascribed mainly to human-made chemicals, notably chlorofluorocarbons (CFCs). And during the 1990s computer modelling demonstrated that ongoing climate change was accelerated by human activities such as deforestation and burning. In some cases, monitoring of such problems served to condemn industrial developments that had been made possible by scientific knowledge. More frequently, however, the economic momentum of those industries encouraged criticism of the scientific evidence.

Criticism also extended from the physical sciences to biology and medicine. The case of thalidomide is typical of rising public concern about scientific medicine. Thalidomide, a drug synthesized by the German company Chemie Grunenthal, was first marketed in 1957 for 'safe, sound sleep'. Soon prescribed to prevent morning sickness, it was discovered some three years later to cause serious birth defects including deafness, blindness and disfigurement of the limbs. Some ten thousand 'Thalidomide babies' were born around the world, many not surviving to adulthood. The distributing companies, doctors and governments acted swiftly to withdraw the medicine, but the case challenged the prevailing public confidence that had been engendered by the success of penicillin (1942), the range of antibiotics that followed and the Salk polio vaccine (1955).

As medicine investigated a growing range of therapies after the war, incidents of failure inevitably accompanied the reported successes. Acclaimed medical successes such as the iron lung (1930s), kidney transplants and medical ultrasound (1950s) and portable pacemakers (1960s) were contrasted by highly publicized heart transplants from 1967, which for over a decade provided only brief extensions to life. British health concerns have more recently focused on food and medications. Government publicity of salmonella bacteria in eggs in 1988 highlighted food safety and scientific advice to government. So-called Bovine Spongiform Encephalitis (BSE, or Mad Cow Disease), a degenerative disease of the spinal cord, was found in the late 1980s to be transmitted by prevailing feeding practices and to produce a variant of Creutzfeldt-Jakob Disease (vCJD) in humans. From 1998, public fears in Britain of a link between the mumps-measles-rubella (MMR) vaccine and childhood autism led to significant drops in childhood immunization. Public protests concerning genetically engineered foods during the late 1990s led to legislation preventing their sale in several countries. And the 2001 outbreak of foot-and-mouth disease, a viral disease of cattle and pigs, led to widespread culling of stocks in Britain and criticism of the measures recommended by scientific advisors – in notable distinction to French policies on how the disease should be handled. Collectively, such incidents reflected a growing mistrust but significant national differences at the turn of the century about the management of science by government and industry.

This survey of perceived failings nevertheless suggests multiple causes. Decisions to proceed with nuclear weapons can be attributed not merely to key scientists such as Edward Teller (1908–2003) but to politicians and funders; the tragedy of thalidomide to inadequate testing protocols and complaisant prescribing physicians; Chernobyl to inattentive operators, deviation from procedures and inadequate design safeguards; and large-scale pollution to science-based industries but also the economic system in which they operate. It might even be argued that late twentieth century technological traumas are due in part to a cultural acceptance of a form of progress that does not adequately consider the full repercussions of change. How precisely can we attribute blame for such occurrences, and how much power do we, as individuals or as a society, have to avoid them?

Twenty-first century transitions

This chapter began by setting the scene at the turn of the twentieth century. It ends in the early decades of the twenty-first century. Can anything be said about the present century, based on the trajectory of past science? Historians avoid making predictions for good reason: sanguine scientific forecasts have often proved inaccurate. New discoveries, new entities and new perspectives unexpectedly changed the terrain of twentieth-century science. Even so, we are not completely blind.

Historians of modern science and technology have revealed the close integration of scientific activities with economics, politics and culture. These forces provide an effective mechanism for directing human desires and defining goals. And, as explored by the historian of technology Thomas P. Hughes (1929–), they can impose a *technological momentum* that resists deviation from a course for decades or longer. From the viewpoint of the present and with an awareness of such momentum, some features of twenty-first century science can be sketched.

The power of genetics – first defined as a subject only a century ago – has been exploited at an accelerating pace in recent decades. Genetic engineering, despite its name, has become a model for a new form of scientific enterprise. Intimately dependent on fundamental scientific discovery, genetic technologies have provided fertile avenues for further research and application. The field already offers some of the powers dreamt of by Francis Bacon, and aspires to many more. DNA profiling allows evolutionary trees to be traced, and has also become a crucial tool for forensic analysis in law enforcement. Cautious steps have been taken in altering and combining genetic material to create new crops (e.g. the genetically modified Flavr-Savr tomato, 1992) and animals (e.g. Dolly the cloned sheep (1996), part of a programme to 'design' animals that produce proteins of human benefit). The Human Genome Project, funded by the government in the UK and by private industry in the USA, yielded a complete genome map in 2003. The continuing research has provided a cornucopia of data for further exploration.

The medical, social and economic potential of such research is widely appreciated. The perceived benefits for distinct audiences have already fostered considerable momentum. GM foods could multiply food production, and raise profits for biotechnology and chemical companies such as Monsanto. Genetic engineering could provide increasingly powerful tools to define, detect, avoid or repair genetically determined illnesses, a power of great importance to health care systems and their funders. And individuals could be empowered by the ability to alter or enhance their bodies or those of their future children. Collectively, these possibilities offer a seductive attraction that, over the foreseeable future, societies may be unlikely to resist.

The deep embedding of science in modern culture is not unique to genetic engineering. Information technology, a generation before biotechnology, grew with profound links to fundamental science, military and business goals and, eventually, wider social interactions. Digital computers were a product of the Second World War intended to speed calculations such as ballistics and code-breaking. Post-war development was supported by military funding, such as the American military's SAGE system developed by International Business Machines (IBM) during the late 1950s to monitor and control a real-time missile defence system. Commercial applications were also pioneered, such as the LEO 1 business computer developed by the British catering company J. Lyons. At first, scientific uses were a by-product. By the late 1960s, though, computer calculations were increasingly used as a new part of scientific methodology. They were able to explore possibilities beyond experiment. They increasingly supported, and then replaced, the elegant but simplistic mathematical derivations that had been the gold standard for physicists. So the scientific effects of computers were subtle but deep. Computer modelling was no longer a crude adjunct to science; it could extend scientists' vision. But modelling has shifted the unstable balance between empiricism (i.e. direct observation and experiment) and theory. The growing power of computer modelling reveals new territory to be explored, and new complexities to be tackled.

Current examples are the sciences of meteorology and climate change. The first weather satellites (such as NASA's Tiros 1 in 1960) improved on the nineteenth-century tradition of collecting data from individual weather stations. The mere task of tracking weather systems and making short-term predictions (no more than a few days with any reliability) was relegated to computers from the early 1960s. Performing the repetitive calculations needed for fine-grained predictions has become increasingly practicable, but still demands the fastest available computers. The detailed understanding of global weather systems demands computers as a scientific tool.

Even more complex are climatic changes. The earth's climate depends on long-appreciated and obvious factors such as solar heating, the insulating properties of clouds and the moderating effects of humidity. Many more factors, operating at a wide range of scales, are significant though. The chemistry of the atmosphere, oceans and soils are intimately involved: sunlight and temperature change liberate different proportions of chemical species which can migrate via weather systems and interact in unanticipated ways. The biochemistry of little-recognized but abundant life forms can significantly alter planetary chemistry. While photosynthesis was studied from the nineteenth century, the importance of plants and microorganisms in controlling planetary carbon dioxide levels has only recently been the subject of urgent research. Appreciating the inherent complexity of the global climate requires computer simulation: such models provide a view unattainable by small-scale experimentation or theoretical overviews.

Even a handful of physical, chemical or biological properties can be unintuitive to understand. During the early 1980s, James Lovelock's (1919–) 'Gaia' hypothesis had a strong influence on understandings of planetary scale interactions. His first persuasive 'thought experiment' was a simple computer simulation that he dubbed 'Daisyworld'. A simulation now accessible from a number of websites, Daisyworld simulated a simple planet populated only with idealized plants that came in two varieties – dark and light-coloured. The model demonstrated that the plant population could regulate the planetary temperature to suit their propagation. While straightforward to programme and demonstrate, his simulation was counter-intuitive. It also hinted at metaphysical dimensions by suggesting that such simple life forms, or even the planet itself, were in some sense sentient, or that a hidden purpose or plan was revealed by their effects (a claim known as *teleology*).

Teleology: the study of pre-existing design, purpose or goals as the explanation for observed characteristics of the natural world.

Lovelock himself was vociferous in rejecting such interpretations, and stressed that his simple mathematics showed little-appreciated feedback processes following natural laws. As successive models were elaborated to make them a closer match to known physical, chemical, biological and meteorological relationships, the essence of Lovelock's findings was confirmed, and more subtle interdependences were revealed. Climate change research is an example of the scientific deduction and social urgency of the new sciences of complexity.

Such examples of current fields illustrate that twenty-first century science is likely to be different in pace, subject area and methodology than past science. In very general terms, this is consistent with past history: every century of science has had this character, and it would be naïve to suggest that the present century's investigations will follow the same trajectory as those of the past century. All of this suggests an expanding field: future historians of science will have ample new territory to explore and fresh interpretations to pursue.

6 More than dead white European gentlemen?

This chapter departs from our chronological trajectory to re-examine our direction. It provides a second layer of discussion, challenging the approach we've taken and the assumptions that may underlie it. The chapter is meant to foreground the reality of writing history, which is that we select – consciously or unconsciously – the subjects that interest or concern us most directly. The chapter does not need to justify the topics that historians explore because, after all, interests are individualistic and changeable, as the wide range and diverse styles of history writing reveal. We must explain, though, why some subjects have long been neglected by historians of science, and what this tells us about the field and, more generally, about our culture.

There is a reason for addressing this topic at this point in the book. The previous chapter surveyed the history of twentieth-century science. Towards the end of that century, approaches to the history of science were increasingly reassessed by a growing number of historians and others. And this questioning has so deeply influenced scholars that it has altered the discipline of history of science, a topic explored more directly in the final chapter.

A tale of two genders

Women, as a group or as individuals, have been little discussed in this book. An explanation – or even defence – of this omission is contentious, and would have been differently framed in other times and places. I will recount those changing views and the effects they had on the three distinct aspects of gender relations in science: women as seen by science; women as practitioners of science; and (in the next section) science as seen by women.

Have women simply been written out of the story? This explanation, perhaps hinting at an unsustainable conspiracy theory, in fact has considerable merit. We cannot recover the details of those overlooked or forgotten by history, but can at least evaluate the female natural philosophers and scientists who created a sustained impact. Among them were the examples of astronomer Caroline Herschel (1750–1848) and Ada Lovelace (1815–1852), conceiver of programming for the first mechanical computer. In the twentieth century – the period of professional science – the task becomes easier: physicist Marie Curie (1867–1934), discoverer of radium and perhaps the most famous female scientist, is commonly recognized; others include Lise Meitner (1878–1968), known for her work in atomic physics, and Rosalind Franklin (1920–1958), whose x-ray determinations revealed the structure of DNA.

Even these extraordinary individuals hint at difficulties of recognition for women, though. Caroline Herschel is remembered in the context of her more famous brother William, with whom she developed mathematical astronomy, and after whose death she continued to compile star catalogues. Ada Lovelace is associated in the historical record with Charles Babbage and his *difference engine*, the first programmable computer. Marie Curie, although achieving fame in the popular press, found full scientific recognition elusive. The first woman in France to obtain a doctorate and twice winner of the Nobel Prize, she was nevertheless denied admission to the all-male *Académie Française.* She became the first female professor at the Sorbonne, but only by assuming the chair vacated by her husband, Pierre, upon his early death. Only as recently as 1995 were her remains moved to the Panthéon, in Paris, alongside her husband. Lise Meitner, an Austrian atomic physicist who, along with her nephew and physicist Otto Frisch, inferred the existence of nuclear fission, was similarly overlooked. Her colleagues, German physicists Otto Hahn and Fritz Strassman, won the Nobel Prize for the experimental work in 1944. Meitner's experience was in part due to her Jewish background, but also to her gender: when she had given a lecture on cosmic physics in 1922, a Berlin newspaper reported it as 'cosmetic physics'. And, like Meitner, Rosalind Franklin's seminal work on DNA failed to obtain the recognition of the Nobel Prize, which went to Francis Crick, James Watson and Maurice Wilkins.

Each of these female scientists, to varying degrees, found herself in a subordinate role in her social milieu or institution. Each was categorized as assisting or furthering the work of a male colleague, despite accomplishing achievements that were distinctly her own. And, for a variety of reasons, each found less recognition by the scientific community for her work than that of her male associate. As historian Londa Schiebinger (1952–) has argued, their many less prominent female colleagues served as 'invisible assistants' in the scientific domain.

The invisibility of women in science is in part, then, attributable to being unrecognized or written out. This cultural effect amounted to more than merely the biases of journal editors, award committees and historians, though. Science was long represented as a mirror of masculinity by those practising it, and widely accepted in popular culture. Francis Bacon, writing in the sixteenth century, had a profound effect on the thinking of many men of science over the following century, and strongly represented the goals of the Scientific Revolution. For Bacon, science was intrinsically gendered, a male preserve. As encountered in chapter 2, Bacon saw science not as a calm philosophical activity but as an active manipulation and conquering of the natural world (itself traditionally portrayed as a feminine essence). This had associations extending back to Aristotle's notions of the masculine spirit as hot and active and the female as cold and sluggish, and the more general Greek view of women as imperfect men. Men of science were meant to prise out the secrets of nature – and to bend nature to human purposes. This virile intellectual activity was widely accepted as a masculine trait by his contemporaries. Culture defined refined gentlewomen as being passive receivers rather than as initiators.

This could be aimed at targets beyond women as a sex, though. Bacon employed his figurative language not only to reinvigorate science but to separate it from the past: he cast Aristotle's work as feminine, being passive and weak. His new science was a programme laden with power relations and sexuality.

A century later, in the publisher's introduction to a 1664 treatise by Robert Hooke entitled *Experiments and Considerations Touching Colours*, Henry Oldenburg, Secretary of the Royal Society, argued that its goal was to 'raise a Masculine Philosophy … whereby the mind of man may be ennobled with the knowledge of solid truths'. Gender remained an effective put-down. His English contemporaries emphasized their science as active and productive, unlike the more 'feminine' arts prevailing on the Continent.

The currency of such views through the Victorian period is illustrated by *Princess Ida* (1884), a comic opera by Gilbert & Sullivan based on the premise that a women's college is a patently ridiculous idea. (The opera has already been introduced in the context of Darwinism.) It was based on a satirical poem by Alfred Lord Tennyson written at the opening of Queen's

College in 1848, a girls' school. (Not until 1869 was Girton, the first women's college, established at Cambridge University.) In one song enumerating the impossible subjects that might fill such a college's curriculum, ranging from 'getting sunbeams from cucumbers' (recalling Jonathan Swift's land of Laputa) to 'finding perpetual motion' (an aim proven pointless by scientists of the day), the stanza underlines the theme:

> These are the phenomena
> That every pretty domina
> Is hoping at her University we shall see!

In short, a woman practising science would be not only inept at identifying suitable problems and solutions but would also seek to dominate her male counterparts, an unnatural abomination!

Behind these assertions were jostling claims about the inherent abilities of men and women. As Bacon and Oldenburg had suggested, women were commonly identified as having the wrong emotional nature to pursue science. Later medical men suggested that female intellectual capacity was also suspect, theorizing that the physiological demands of menstruation, childbirth and nurturing of children chronically limited their abilities to think and plan abstractly. Even worse, such inherent weaknesses were thought to make women, as a sex, peculiarly susceptible to illnesses such as hysteria. Here again, Greek roots can be traced: Greek medicine explained hysteria as an uncontrollable illness caused literally by movement of the womb within the body (the root *hyster* is used for both). While Victorian explanations had ruled out this physiological cause, they continued to attribute the illness to women alone. Sigmund Freud (1856–1939) recast hysteria as primarily a woman's illness caused by psychological trauma. The pejorative dimension of the term 'hysterical' applied to women can be suggested by considering the adoption of the term 'testerical' in relation to men. Subsequent medical men were able to employ both physiological and psychological explanations by suggesting that the prevalence of hysteria had increased owing to the social pressures of female emancipation and women in the workforce during the First World War. In any case, this 'women's illness' ceased to exist formally in the mid twentieth century, when it was removed from standard psychiatric categories of disease entities.

Essentialism: the belief that certain characteristics or behaviours are fixed and unchangeable.

These judgments of female competence, often supported by prevailing scientific views, go some way towards explaining the small numbers of women contributing to science. They have remained significant factors until the recent past. Take, for example, membership in professional societies of science and engineering. The Institution of Chemical Engineers, originating in the 1920s, found itself short of members during the rapid expansion of the chemical industry after the Second World War. Unlike more conservative technical professions, the Institution sought female members, and indeed achieved the largest proportion of women of any comparable society. Even so, this was a minor change until the 1980s, with most women still in the more junior membership classes. At the present rate, women would represent half of the membership by the mid-twenty-first century, still a generation or two away.

The gradual entry of women into the technical workforce has social and cultural origins that would have been familiar to nineteenth-century men of science. When the Institution of Chemical Engineers surveyed employers in 1958 about their preferences for female engineers, half of the firms stated that they would not employ a woman. The remainder preferred female engineers for sales or research, but blamed their decisions on

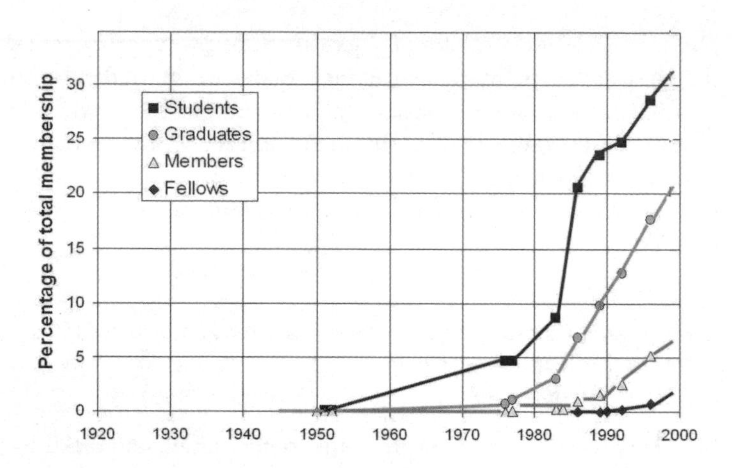

Figure 12 Female membership in the Institution of Chemical Engineers, 1922–2000 (S. Johnston)

women's faults in personality and management ability, the likelihood of marrying and leaving employment, and indecisive career plans. The Institution itself recorded marital status for its female, but not male, members, and advised universities not to fill many places with female students owing to their lower expectations of useful careers.

The documenting of such cultural attitudes is a key part of explaining gender differences in science, and begs the historian to go one step further. What, she can ask, would an ungendered, or even female-oriented, science be like?

Class and causation

An equally obvious line of discussion concerns the importance of social position in creating science. Our survey has often encountered influential practitioners who were members of a privileged elite. Many of the best-known names – Galileo,

Newton and Darwin, for instance – were well-born, affluent or otherwise high-status gentlemen. Their prominence suggests a 'top-down' understanding, in which a leisured and well-resourced class extended scientific knowledge based on intellectual refinements.

On the other hand, we have also focused periodically on 'bottom-up' science, too: the extension of knowledge by the innovations and manual experimental skills of working-class practitioners. The experiments of the Royal Society were made possible by the skills and technological innovations of instrument makers, for example, and the close working relationship of Joseph Black and James Watt simultaneously advanced thermodynamics and efficient steam engines. Even so, most of the artisans mentioned were those who shifted economic class to become socially visible. Michael Faraday, the prominent nineteenth-century electrical experimenter, had started life as a book-binder's apprentice and ended as a reputed scientist and government consultant. James Joule, son of a wealthy industrialist, employed the skills of his trade to pursue experimental science. And Watt himself went from instrument maker to company magnate. Even as brief a survey as this book, then, cannot reduce the account to two competing classes or alternate routes to knowledge, or to neatly segregate 'ivory tower' science from a simple economic account of technological innovation. Historians continue to grapple with these multiple contributions and differing accounts.

Nevertheless, the workers are more difficult to identify for some of the same reasons that female practitioners have been. As another form of invisible assistant, craftsmen and artisans created products that were awkward to transmit as their own to subsequent generations. Popular inventions and successful skills become rapidly anonymised and shared, while experimental findings and conceptual frameworks often retained their provenance. Disseminated by practise rather than by publication, craft skills relied on unacknowledged workers of relatively low social esteem.

Here, again, we can pose questions for ourselves. How faulty are our understandings of history of science because of a bias towards high-status individuals and their cultural products? As discussed in chapter 7, historians have become aware of the amount of catching-up to do.

Non-Western perspectives

Our perspective can be extended to further acknowledge under-represented communities. Is science a universal activity, or limited to certain cultures? Chapter 1 argued that some of its attributes, at least, seem ubiquitous in human societies. On the other hand, relatively little space has been devoted to non-Western societies, largely because of a relative lack of scholarship by Western historians. (Such voids urgently suggest future directions for research.) Moreover, this originally Mediterranean approach to knowledge has spread to influence global perspectives in the contemporary world, making it relevant to ever-wider audiences.

The question of science's universality has yielded mixed responses. On the one hand, it seemed apparent to most nineteenth-century observers that certain societies had progressed in civilization, and that this supported prevailing ideas of cultural and even racial hierarchy. (Needless to say, such claims are now highly disputed.) On the other hand, men of science had long argued that intellectual advance was an objective enterprise, and one that could be delayed or tainted only temporarily by social factors. Until recently, and still the dominant view today, histories of science have supported the claim that the Western trajectory of science represents the most effective route to new, reliable knowledge. To consider other contexts, according to this view, would be, at best, a scenic diversion. Since the 1960s, the limitations of this Western-centred viewpoint have increasingly been recognized. I will say more about those changing stances in chapter 7, but can illustrate them with a few historical cases.

The Western bias in interpreting other cultures is well illustrated by the case of China. Scholars during the late nineteenth and early twentieth centuries, when China was successively partitioned, occupied and riven by civil war, had frequently dismissed the culture as intellectually unprogressive. Such views were challenged increasingly from the 1940s by the lifelong studies of one Western scientist and historian, Joseph Needham (1900–1995) at Cambridge. Through his ultimately unfinished series of texts *Science and Civilization in China* and others, Needham's principal theme was to document China's achievements in their social, cultural and political context, and to explain the country's divergence from the Western trajectory of development. China's earlier adoption of powerful methods, technologies and concepts – notably celestial observations, printing, the compass and gunpowder – did not lead to a similar expansion of scientific activity. Needham argued that the religious environment of Confucianism and Taoism had a strong influence on the pursuit of science. His analysis illustrates the importance of social context not just in interpreting the past but also in shaping how we view it. More recently, the rapid economic growth of China in the early twenty-first century has been accompanied by re-evaluations of the country's entire history. No longer interpreted as a stultified and uncompetitive culture, China is now more often viewed as remarkably enduring and adaptive, with the rise and fall of scientific activity being attributed more directly to changing economic circumstances.

Such biases concerning non-Western indigenous cultures can be distinguished from another and more subtle form of

Eurocentrism. It argues that colonial societies, too, have been viewed as lower in the scientific hierarchy because they have been judged by standards defined by their mother-countries. For example, colonies and territories founded by Spain and Portugal (e.g. the countries of Latin America), Britain (e.g. India, Canada, Australia) and France (e.g. Quebec and Indochina) attracted settlers from the mother-countries and developed administrations and cultures based on them, often with an integrated but subordinate status for indigenous peoples. According to such models, overseas territories represented an extension of European practices and values. Indeed, European missionaries often operated in parallel with exploration by geographers and naturalists and with economic exploitation by organizations such as the East India Company (1600) and Hudson's Bay Company (1670). Local scientific investigation of phenomena in these regions, and technological solutions to local problems using local resources, would be evaluated according to European criteria and, more often than not, found wanting. Colonization thus imposed a dual hierarchy: native peoples subordinate to the colonizers, but colonial societies as subordinate to their mother-countries. Not only social status but intellectual status, too, was at play. The value of locally produced knowledge, by either the colonizers or the colonized, was often found wanting. As a result, these territories tended to lose indigenous methods of acquiring knowledge but were disadvantaged in absorbing new methods on their own.

This viewing of colonial science as marginal and inferior has been dubbed *peripheral science*. Historians and sociologists in recent decades have argued that science cannot be understood as the export of knowledge from metropolis to periphery. Instead, it is necessary to understand separate locations and societies on their own terms. This diminishes claims about science being a universal activity best practised in the European style. It replaces the Eurocentric (or Western-centred) view by investigations that analyze the local context of knowledge, discovery and application.

This liberation of viewpoint creates challenges of interpretation. Most importantly, it discourages scholars from a judgmental or normative approach to the history of science. For instance, the rhetoric of colonization traditionally has emphasized the improvement to indigenous cultures provided by scientific achievements and technological solutions. But examples of cultural trade in the opposite direction can readily be discovered. A wide range of foods, cooking techniques and medical practices from South America and Asia have also been adopted in the West. More generally, a strong case can be made about colonizers' adaptation to their local circumstances by drawing upon indigenous knowledge.

The conventional colonization claims may also downplay negative consequences. The European colonization of North America between 1500 and 1900 shared knowledge and benefits inequitably with native peoples. Diseases imported from Europe such as measles, smallpox, chicken pox and influenza repeatedly decimated local populations, and the later introduction of large-scale agriculture was at the expense of native American ecology and hunting cultures. More recent examples of the mixed effects of inter-cultural relations include genetically modified seed for third-world countries: some varieties, while promising better crop yields, are not self-fertile, making the indigenous buyers dependent on annual purchases from the seed company.

It's how you tell the story ...

Discussion of national and gender dimensions of the history of science reveals how the subject increasingly is approached. Can we explain the changing involvement of women solely through examining intellectual history? Most historians today would answer: probably not. Did a masculine vision of science shape its direction, its achievements and its evaluation? To a degree, almost certainly so. Can such analysis provide insights of value for cultural studies, educational curricula and science policy? I think so.

From the cases mentioned in this book, it will be apparent that the history of science has the power to inspire, defend and reveal. The study of how history is written, known as *historiography*, has a particular value in history of science. (It is also a key topic in general historical studies and is closely related to themes in media studies, notably concerning how messages are framed and constructed to suit goals and audiences.) Some approaches and examples will illustrate its influence on ideas and culture.

> **Historiography**: the study of how history is written, or the craft of writing it.

An approach long derided in general history, but with traces still surviving in some history of science, is the wholly positive portrayal of its influential figures. The style is known as *hagiography* after the biographies of saints written in the Middle Ages. Some scientific biographies have been constructed very nearly as stories about individuals imbued with sacred qualities. They commit a number of faux pas. Such stories are generally uncritical about the certainty or judgment of historical facts. They downplay, or even omit to mention, the complex social factors while emphasizing clarity of intellectual achievement. They adopt a respectful, sometimes reverential, tone concerning the subject's character. Most describe the scientific career as a linear road of progress, sometimes with external barriers to be triumphantly overcome. Some even hint at the realization of a hidden purpose or destiny, an expression of teleology. In short, they simplify complex life stories to construct a narrative tale of success and even moral example.

Popular examples are often rather old ones, because this style of historiography has been out of fashion for a few generations. They include the cinematic biographies of Louis Pasteur (1935) and Marie Curie (1943). This admiring hagiographical treatment can also be extended to more abstract aspects of science, though. Paul Erlich's (1854–1915) discovery of Salvarsan, a treatment for syphilis, was immortalized in *The Magic Bullet* (1940). In that film, the reverence is associated not just with the saintly aura of Erlich, but extends to the near-miraculous medical treatments and the power of the scientific method itself. Science, it strongly suggests, is a moral good.

Hagiography is not merely a side-effect of science popularization, though. It is amenable to views of scientists as figures who rise above, or set aside, personality in pursuit of a greater goal. This has a close affinity with a framing of the history of science as merely intellectual history, according to which a sufficiently pure or uncompromised individual can uncover new objective knowledge. And sometimes, hagiographic writing can have the more mundane aim of assuring a posthumous reputation for scientists whom their colleagues think have been underappreciated. Lewis Campbell's *The Life of James Clerk Maxwell* (1882) is a typical example.

Where hagiographic scientific biographies have deviated from the reverential model, they have often begun with picturesque anecdotes that illustrate a life beyond science, but enriched and rounded. Campbell's biography, for example, discussed Clerk Maxwell's unpublished poetry, humour and devotion to his wife. *The Life of William Thomson: Baron Kelvin of Largs* (1910) alluded to Lord Kelvin's interest in family, music and – important in the period after Darwin's theory of evolution – religious convictions.

But hagiography can produce the opposite result, too: a depiction of an unworldly, objective and pure scientist who may seem inhuman, impersonal and isolated. These two sides became closer through the twentieth century. As a result, science and scientists could be represented or understood as simultaneously powerful and remote. Angels and devils had the same origin, after all.

Following an increasingly critical evaluation of science from the 1960s, such scientific biographies became unfashionable. A generation later, however, biography was redeemed by critical studies such as Stillman Drake's *Galileo* (1980) and *Darwin* (1992) by Adrian Desmond and James Moore. These carefully interrelate the lives, work and ideas of their subjects rather than dividing the account into internal and external components (for more on this division, see below). More recently still, spates of popular scientific biographies have been published that adopt exposé-style revelations to overturn widely held convictions. They amount to a delayed reaction to the mid-century hagiographies that dominated scientific life stories. A variety of approaches survives to influence scientific biographies today, as suggested by some titles intended for semi-popular readerships: *In Albert's Shadow – the Life and Letters of Mileva Maric, Einstein's First Wife* (2003), by Milan Popovic; *John Von Neumann: The Scientific Genius Who Pioneered the Modern Computer, Game Theory, Nuclear Deterrence, and Much More* (2000), by Norman MacRae; *Neils Bohr's Times – in Physics, Philosophy and Polity* (1994), by Abraham Pais; *Gerhard Herzberg – an Illustrious Career in Science* (2002), written by colleague Boris Stoicheff.

Digging deeper into getting it wrong

Gender and biography are illustrations of minefields for historians of science. Such examples introduce other regions of historiography that demand equal caution. History of science can, intentionally or not, channel hidden or unconscious viewpoints. It may even support overt propaganda.

This is not a new or radical view, but was first voiced in the early 1930s by the historian Herbert Butterfield. Butterfield described what he called *Whig* (or *whiggish*) *history*. His criticism focused on a popular style of history writing that consciously or unconsciously supported 'the winning side'.

Whiggism: a biased approach to history-writing that interprets past events according to present-day standards and perspectives, and which limits historical attention to historical events that have been interpreted as contributing to present-day science.

Butterfield's prime examples were British histories written in the nineteenth century. By that time, Britain was an imperial power at its greatest extent. Histories of the nation could represent this as a natural culmination. The causes might be attributed to various factors: political refinement (e.g. culminating in the Whig party), enlightened morals or, indeed, intellectual advance and technological application. Butterfield's point was that this was to a degree a mirage created by the selective interpretation of the facts. It also threatened to be self-serving; such histories invariably supported the status quo. This idea has an even longer lineage in literature. It is a theme of Voltaire's satire *Candide* (1759), in which the central character invariably and unconsciously contrives to interpret events in such a way as to demonstrate that he lives in 'the best of all possible worlds', and is also the premise of Rudyard Kipling's *Just So Stories* (1902) that facetiously trace the unsuspected origins of animal traits. The philosophical criticism of this approach is that it is *teleological*, that is, it hints at a purpose for things being the way they are.

How can whiggish history be identified? Butterfield suggested that a history that vaunted the present as being the highest and most perfect in human history was at least suspect.

Others have argued that such an interpretation of history is more likely if the historian adopts a perspective that sees the present day as normal and other periods as imperfect or unrefined.

This *presentist* perspective encourages us to accept a series of conclusions in a snowball effect. First, it suggests strongly that we live in an age when the important things have never been better. (Those 'important things' may be defined differently by different observers.) Second, it suggests that if the past was poorer in these respects, then the future is likely to be better. And third, it may suggest that we can identify the factors that account for the improvement in those 'important things' and apply them to *design* a better future.

Presentism: the analysis of past events in terms of present-day perspectives and ideas.

This may raise alarm bells for some readers – isn't science about just that? Surely scientific knowledge today is the best it has ever been, and will progress unproblematically as long as we identify and continue to follow the best possible method?

To a degree, most would identify aspects of science in these assumptions. Philosophers have argued that we have refined scientific methodology, and become more conscious of the successful methods. Historians can chart many variables linked with scientific knowledge that have improved over time. Historians of the late twentieth century have argued, though, that these characteristics are not as clear-cut and general as once believed, and may to some extent be illusory.

William Whewell could claim unproblematically in the early nineteenth century that humankind, 'from the time of its creation, has been travelling onwards in pursuit of truth'. The high-point of these assumptions came with the career of his French contemporary, Auguste Comte (1798–1857), introduced in chapter 4. Comte promoted his concept of *positivism*, which argued for the progress of civilization and for a methodology of assuring more rapid progress. Comte argued that human history was a chronicle of progress. This progress was in several forms: religions had 'progressed' in three stages, from animism to polytheism to monotheism. Religious ideas themselves, he said, had withered as rational explanations grew. And within the rational modes of thought – the sciences – he could discern a similar steady upwards slope. Astronomy and mathematics, the oldest sciences, were the most quantitative and objective. Younger sciences such as chemistry were on the right course, but not yet sufficiently mathematical. And the youngest would-be sciences – such as the field of sociology, for which he coined the name – could aspire to follow the older models. Progress, in short, was inherent in human societies. Comte claimed that human knowledge followed an invariable chronological sequence. This was clearly a hierarchical and Eurocentric vision of humanity, and one that supported the general whiggish approach to historical interpretation.

For generations of historians, these conclusions appeared to underpin twentieth-century improvement, too. Scientific activity was increasingly organized. Many scientific attributes unquestionably improved: the number of scientific journals, the number of papers published, and the number of scientists themselves. New phenomena were discovered and ever more coherently explained. Some technologies benefited immeasurably: the field of electrical machines, and then electronics and computers, blossomed. Historians have more recently argued, however, that this is a myopic and 'fundamentalist' view. It focuses attention on particularly active sub-fields in science (e.g. nuclear physics) while ignoring others that showed less remarkable expansion (e.g. photometry). It spotlights some forms of technological change (e.g. expansion of motor transport or the capacity of computer hard disks) while ignoring others (such as climate change and the disposal of nuclear waste). It usually ignores most social factors, and indeed encourages the idea that knowledge always has a net positive social effect.

Belgian historian George Sarton (1884–1956) represented his generation in claiming that science was the only human activity that was obviously cumulative and progressive. In his inaugural lecture as the first historian of science at Harvard University in 1935, he argued that the history of science was 'the only history that can illustrate the progress of mankind'. Even opponents of positivism, such as Alexandre Koyré (1892–1964), stressed the autonomous and progressive advance of *theoretical* ideas, if not experimental facts. On the other hand, there were hints of dissent. An outsider who made a strong impact was Soviet physicist Boris Hessen. His 1931 paper 'the socio-economic roots of Newton's *Principia*' argued that history of science should focus on social context and economic circumstances (incidentally, Hessen was executed five years later as an enemy of the State).

Koyré shunned such cultural and social factors in the development of science. The counter-side to this century-long belief in assured progress began more assuredly in the 1960s. Two years before Koyré's death the publication of Thomas Kuhn's *Structure of Scientific Revolutions* challenged this form of idealism. As touched upon in chapter three, he argued that the development of scientific knowledge was not a continuum but rather a series of profound shifts, which were the result of confrontations between awkward experimental facts and paradigms, or theoretical frameworks. More subversively, Kuhn's work implied that such perceptual shifts often relied on subtle social factors such as the status and institutional affiliations of key opponents. These factors were teased out and pursued by subsequent philosophers. Significantly, though, Kuhn's book, despite its immense influence on the field, had no explicit discussion of the social domain.

Further case studies explored histories of scientific development that incorporated their social and economic context. They suggested that scientific directions were not always followed because of the clear superiority of intellectual concepts. Social factors could, and did, have an effect on the history of science. A first wave of cautious studies suggested that this could be broken down into two contributions: an internal and external effect.

Internal history of science was the type that had been pursued for over a century. The logical chronology of scientific discovery and theorization could be tracked. Philosophers found such information essential to understand the evolving methodologies of science. Internal history also could be pursued readily by scientific experts in the field. As we have seen, however, this could at times bias their accounts towards presentist and positivist history. It could even lead to a particularly contentious variety – 'triumphalist' history, in which current science could be promoted or even propagandistically claimed to support government policies.

The newer form of external history of science was more often pursued by outsiders such as the sociologist of science Robert K. Merton. In some respects, Merton was more of a renegade than Kuhn: as a physicist, Kuhn had developed in an environment attuned to the philosophical orthodoxy of science, even if his later work challenged it. Merton, by contrast, focused on the sociological dimensions Kuhn had not broached. Merton argued that the expansion of scientific knowledge required particular social conditions to thrive. These conditions included the sharing of information (a 'norm' he referred to as communism), a lack of personal attachment (disinterestedness) to scientific claims, and rigorous investigation of claims (scepticism) without attributing them to individuals ('universalism'). According to Merton's understanding, science could be diverted or corrupted in environments that did not foster and support these social criteria. So long as the right conditions were present, though, scientific knowledge could be understood by an internalistic examination of its methodology. Later historians and sociologists of science have focused on the particular social and cultural factors influencing scientific history, and tended to avoid the fine details of intellectual development. Despite early criticism by practising scientists who argued that their 'insiders' accounts' were crucial to understanding their results, this 'external' approach, at least as a first approximation, considerably broadened understandings of how science worked.

A subsequent generation of historians has sought to reconcile the internal and external accounts of science by probing deeper. In survey texts like this one, it can be difficult to avoid the seduction of 'grand narratives' that explain large-scale events with sweeping explanations and inevitable hidden biases. Surveys – at least in written form that can be picked apart by careful scholars – have become unfashionable. By contrast, and like most developing disciplines, scholarship has focused increasingly on the deep and narrow. So-called 'microstudies' try to address the full context of historical episodes. While this can reveal satisfying explanations it also obscures the field for non-specialists, unable to see the forest for the trees. While the verifiable detail of microstudies is our professional aim, many historians are tempted to explain more, and aspire to proselytize for their field. The present book – a relatively crude gloss of facts and interpretations – is, I believe, a necessary reaction to our disciplinary goals.

All this professional squabbling can sound rather discouraging and abstract. The crescendo of steady improvement is exhilarating to be part of and stimulating to document. Challenging this aim – if not steady achievement – seems ungrateful at best, and perhaps even motivated by darker intentions. But critical historiography should not be thought of as a pedantic spoiler of enjoyment or as a threat to progress. It can strip off the superficial layers of interpretation of historical events, and reveal more subtle depths of explanation. It can highlight the fertility of human intellectual endeavour and the multiple directions of its offshoots. Historiography empowers us to dig deeper to better understand how science grew to be this way.

7

Science, history and culture: evolving perspectives

This chapter could be entitled 'the present and future of history of science'. As the previous chapter suggests, history of science has increasingly been re-examined 'from the inside'. But not only have historians revisited the subject: so too have philosophers, sociologists and the wider public. Together, these visions of the subject have reworked its goals, its methods and its audiences. The aim in this chapter, then, is to explore behind the scenes to illustrate how scholars have drawn meaning from the history of science.

Philosophers and failure: disputing how science works

From the early nineteenth century, the history of science became closely associated with philosophy, although, as historian Simon Schaffer (1955–) and others have argued, the writing of scientific history also developed from biographies of exemplary practitioners. Both connections were only to be expected. Mediaeval categories had defined scientific knowledge as part of natural philosophy, an understanding traceable to Aristotle; and big names like Aristotle provided authoritative answers. The narrower term *physics* – dropping the philosophical association – was substituted increasingly during the late nineteenth century, but there were some hold-outs. Scottish universities largely retained the term *natural philosophy*, with Glasgow University's department being reorganized to become the Department of Physics and Astronomy only in 1986.

Elaborated by William Whewell, the history of science served as raw material to construct philosophical understandings of the universal nature of knowledge. He categorized branches of scientific knowledge, traced their histories, and coined new words (such as *physicist*, *anode* and *uniformitarianism*) to describe them. Whewell's *History of the Inductive Sciences* (1837) was followed quickly by *The Philosophy of the Inductive Sciences, Founded Upon Their History* (1840). The twinned books argued that the successful advance of scientific knowledge relied on *induction*, the generalization of concepts and laws from particular examples.

Whewell's French contemporary (and rival) Auguste Comte, introduced in chapter 4, also categorized the sciences via a historical survey. Unlike Whewell, Comte suggested that the advance of knowledge could be accelerated by a rigorous methodology that relied only on observable facts. 'Positive' knowledge extended only to what could be experienced directly (empirical evidence). And, as singer Bing Crosby put it, in addition to accentuating the positive, it was equally important to eliminate the negative: theorization about hidden or abstract entities, Comte claimed, was pointless.

Ironically, while the history and philosophy of science became closely allied, practising scientists drew away from both fields around the turn of the twentieth century. Scientists, now increasingly professionalized and goal-oriented, delved into the philosophical justifications of their work less often. Perhaps the last influential scientist-philosophers were Ernst Mach (1838–1916) and Pierre Duhem (1861–1916). A confirmed positivist, Mach argued that scientific laws are merely short-hand for collections of experimental findings. His book *The Economical Nature of Physical Inquiry* (1882) suggested that 'the law always contains less than the fact itself', and promoted the trimming of scientific claims to only what was strongly supported by evidence. This rather fundamentalist approach led him to reject the atomic hypothesis at the turn of the century on the grounds that direct observations were impossible, and the indirect evidence was 'uneconomical'.

French physicist and philosopher Pierre Duhem, on the other hand, investigated the problem of constructing scientific theories. His position, known as the Quine-Duhem thesis or the problem of *under-determination*, is that for any set of data a large number of theories can be convincingly applied. In any experiment then, the results will not provide sufficient evidence to force revision of a theory. (This came up in the discussion of the Michelson-Morley experiment.) This unsettling conclusion raises questions that were reviewed by historians and sociologists of science half a century later: how can empirical evidence be related to theories at all, and (more fundamentally) what confidence can we have in scientific realism – the notion that it is possible to discover the true nature of things?

Philosophers, too, were stimulated (and disturbed) by the elaboration of new scientific theories such as relativity and quantum mechanics. One outcome was the influential *logical positivist* movement which sprang up in 1920s Vienna and came to dominate American philosophy of science into the 1960s. Inspired by Comte's positivism, its members questioned the proliferation of concepts that could be only indirectly inferred. How, for example, do we know that 'electrons' and 'energy levels' truly exist? They argued for a more logically based scientific method, and could turn to historical episodes to illustrate how successful science had operated in the past.

While scientists themselves were turning away from philosophy, then, history of science developed in close association with it. Early twentieth-century historians of science emphasized the evolution of concepts and the accumulation of factual knowledge – intellectual history and its philosophical significance. The label 'History and Philosophy of Science' (HPS) identified a successful proto-discipline at prominent institutions such as Cambridge. Part of the reason for the success there can be attributed to its location within the Faculty of Natural Science. Academic affiliations were also tight elsewhere: the History of Science Division at the University of Leeds, for example, was founded in the mid-1950s within its Department of Philosophy.

More frequently, though, Philosophy departments pursued the philosophy of science without explicit links to the history of science. Notable contributors – each with well-established explanations and followers – included Karl Popper (1902–1994) at the London School of Economics and Thomas Kuhn at the University of California, Berkeley. Both rejected the positivist philosophy that was becoming the orthodox explanation for how science works, but for distinct reasons. Popper emphasized insights about the scientific method. Importantly, he demonstrated that scientific theories can never be wholly proven: at best, they are conditionally confirmed by mounting evidence. In essence, this is a critique of induction. He proposed instead the notion of *falsificationism* to explain the advance of knowledge. Facts, he argued, could never be confirmed to be true in general, but they could be proven false. For example, the claim 'all swans are white' cannot be proven true unless we examine every swan that exists or has existed, but it can be proven false merely by finding a single example of a black swan. Swans aside, the same imbalance affects many modern claims: we could prove that UFOs or ghosts exist by capturing just one, but we can never disprove their existence; we may be failing to find them because we are looking in the wrong places or at the wrong times! Science advances, said Popper, by testing falsifiable hypotheses. The remaining set of hypotheses not yet proven false represents our working body of knowledge. Some historians argue that there are few examples of this method in practice.

On the other hand, Kuhn, as we have seen, claimed a different use for historical evidence, demonstrating that it did not support the notion that science accumulated knowledge inexorably; periodically, he showed, there were ruptures of knowledge, and new frameworks of understanding or world views) were constructed. The revolutions discussed in chapter 3 are evidence of such periodic convulsions. Both Popper and Kuhn emphasized the importance of theory-making, and so reduced the stress on fact-collecting and empirical observation.

Imre Lakatos (1922–1974) and Paul Feyerabend (1924–1994), both junior associates of Popper, provided their own distinct approaches to the philosophy of science. Lakatos sought to reconcile the views of Popper and Kuhn. Feyerabend's *Against Method* (1975) argued that science is not a unified body of knowledge with any identifiable universal method; instead, it is an incoherent patchwork of particular techniques and procedures that isolate pockets of knowledge.

As this outline suggests, the history of the philosophy of science has interesting parallels with the history of science itself. A series of contributors (including Whewell, Comte, Popper, Kuhn, Feyerabend and their successors) have reformulated the bases of the subject. As a result, the world views of philosophers of science have been shaken periodically.

Science in the post-modern world

As sketched in chapter 6, the close alliance between the history and philosophy of science was broadened to other disciplines during the late twentieth century. There has been a flourishing of approaches to writing the history of science, and to understanding science itself. These new viewpoints have resulted from studies of science 'from the outside': from other disciplines and even from other belief systems. This gradual process, developing after the Second World War, can be described as waves of reorientation for the history of science. Two such changes of direction have been described as the 'linguistic turn' and the 'social turn'.

The linguistic turn

The phrase 'linguistic turn' refers to a shift in the history of science towards studies of language and discourse – i.e. research into the ways that scientific findings have been described, communicated and perceived. This refocusing of attention towards language began in Humanities subjects (history, literature, cultural and media studies) during the 1960s. These ideas amalgamated developments by philosophers such as Ludwig Wittgenstein (1889–1951) and literary theorists such as Jacques Derrida (1930–2004). Wittgenstein had argued that philosophical concepts are intimately tied up with language. A separate development was the method of *structuralism*, conceived first for the study of linguistics but spreading to other fields during the 1950s. Structuralism seeks to find abstract patterns or structures within social events, and to determine their rules of combination. Anthropologists such as Claude Lévi-Strauss (1908–), for example, sought to discover the 'deep grammar' of societies by studying their rituals, kinship associations and mythology.

By contrast, the next generation of philosophers and critical theorists, particularly in France, developed a critique of structuralism that, logically enough, has been dubbed *post-structuralism*. Derrida, Michel Foucault (1926–1984) and others argued that various radical philosophies that were critical of Western philosophies illustrated the degree to which Western culture itself had defined ways of thought. Post-structuralists argued that the underlying 'structures' of society identified by structuralists are not universally observable characteristics but are, in fact, conditioned or created by culture. As a result, the attempt to apply a scientific approach to social processes is inherently biased. They seek, instead, to understand the world by investigating multiple viewpoints or perspectives. Such a position clearly challenged the interpretations of the history of science that had been developed up to the 1960s. It also challenged the traditionally privileged position held for scientific knowledge itself.

While its origins are diffuse and definitions are spurned, at the heart of the linguistic turn is the conviction that our understanding of the world is strongly filtered and shaped by language. Rhetoric, it is argued, can create a vision of the world by defining terms and corresponding concepts. This highly constrained perspective can limit or distort our perceptions about the natural world. Taking up this approach, historians of science began to study scientific texts in relation to other kinds of discourse, such as religious and political. They sought to discover how scientists' discourse affected the presentation of their findings and how texts had been used to persuade audiences of their explanations. This approach probed the motivations for scientific accomplishments, and related them more clearly to the context of their times and to other varieties of historical study. Science, previously explained mainly by the logic of reasoning, was now more fully explained in terms of ideologies and interests. One might ask, for example, how we can characterize Isaac Newton, Robert Boyle or Louis Pasteur by the books they read and wrote. The attention to scientific rhetoric led naturally to wider interest in historical context that created these texts. It also opened the door to the study of different national settings for science. Studies of articles on science in popular journals, for example, embedded the field more firmly in the scholarship of

Victorian culture, and revealed differing perceptions of science in the countries of Europe and America.

The social turn

As suggested by the linguistic turn, growing numbers of historians of science began to focus on the rhetorical and social factors underlying scientific knowledge. A closely related shift in attention from the 1970s was the so-called 'social turn'. Here, too, the roots of change can be found in the ideas of other disciplines.

Social history, a branch of historical studies that had been growing since the 1950s, called attention to 'history from below'. Its proponents argued that social norms and beliefs could arise from, and be sustained by, the masses rather than from figures of authority. Applied to the history of science, social history focused on audiences and different portions of the public – by class, education, occupation or national origin – rather than on men and women of science. Histories of science could now be devoted to the *reception* of scientific ideas instead of merely to their *creation*. Indeed, social history has more recently extended to *cultural* history of science. This approach focuses on the relationship between science and culture, a subject that makes sense only if the historian appreciates that knowledge may have different cultural expressions or be subject to cultural shaping. For instance, the national preference for the concepts of Descartes in France was diminished after the French translation and commentary of Newton's *Principia* by woman of science Émilie du Châtelet. The fruit of this approach is the broader understanding of science as a social process: a collective human activity fraught with human emotion, motivations and mistakes as well as successes.

A second outcome of the social turn was a new attention to craft skills and artisanal knowledge as drivers of science. As suggested by philosopher Jerome Ravetz (1929–) during the early 1970s, this approach counteracted the traditional interest in intellectual history for the field. Rather than focusing on concepts and their mutation, historians of science increasingly investigated the importance of process skills. Historian Myles Jackson (1964–), for example, has argued that the development of spectroscopy by Joseph von Fraunhofer in early nineteenth-century Germany owed much to his artisanal expertise in precision optics.

But just as post-structuralism had influenced 'the linguistic turn', it provided a more radical slant for the 'social turn' than did social history. The relationship between social activities and scientific findings can be explored one step further: can society not merely constrain or filter our scientific practices, but also shape our scientific beliefs? An example of this possibility is the investigation of N-rays discussed in chapter 5. The N-ray studies and claims of René Blondlot had been conditioned by his working environment. The contemporary discoveries of x-rays and radioactivity, along with state-of-the art practices of measuring the brightness of light and detecting radio waves, had made him and his scientific collaborators receptive to interpretations that might have seemed improbable in other social contexts. The data reported in papers by several laboratories were later judged to be illusory and misguided. On a much larger scale, it can be argued that the Aristotelian understanding of the heavens – the Western orthodoxy for nearly 2,000 years – was supported by prevailing theological ideas and trust of ancient authority.

An extension of such anecdotal cases suggests that, at least sometimes and for certain periods, scientific facts can be *socially constructed*.

Social constructivism: the view that knowledge is a human product that is socially and culturally shaped, rather than being based primarily on discoverable physical reality.

This new approach had at least two clear consequences. First, it brought scholarship in the history of science closer to other fields such as literature, anthropology and sociology. This new cluster of scholarly interests has been described by labels such as *science studies*, *science and technology studies* (STS) or *science-technology-society* (also abbreviated as STS).

The second consequence of these turns has been that they diminished the perceived separation between scientific knowledge and other forms of human belief. Another intellectual current for historians of science during the late twentieth century came once again from philosophy and sociology. So-called *sociology of scientific knowledge* (often referred to by its abbreviation, SSK) was the product of interdisciplinary studies from the 1970s. One of its principal founders, David Bloor (1935–), argues that sociological factors influence all aspects of science, from the selection of problems to funding, from categorization of results to dissemination, and from observation to theory construction. Members of the so-called Edinburgh School (the Science Studies Unit at the University of Edinburgh from the late 1960s) distinguished two versions of this viewpoint. The 'Weak Program', they argued, applies discussion of social factors merely to erroneous beliefs. Thus Blondlot's N-rays would be attributed to the social considerations that flavoured his scientific work, while the successes of his critics would be attributed solely to intellectual factors and rational, objective judgment. As suggested by the pejorative label 'weak', the members of the Edinburgh School favoured a different approach, which they dubbed the 'Strong Program'.

According to the Strong Program, historical, sociological and philosophical investigations of scientific practice should strive to be neutral with respect to our current beliefs about the truth or falsity of claims. This approach, known as *symmetry* or *methodological relativism*, gives equal attention to historical episodes that today are seen as 'successes' and 'failures'. The history of science is thereby broadened to document and analyze not just how we came to hold our present scientific beliefs, but also the numerous blind alleys, failed initiatives and errors of the past. This is not merely a matter of being 'fair' to the collection of historical actors, but also to better understand the strategies and philosophies of knowledge pursued by our forebears. It also extends history of science into the present and, indeed, the future: the scientific practices and strategies of the past inform those of the present, and so historians, anthropologists, philosophers and policy-makers have much to teach each other.

A more radical position of the Strong Program, however, is its commitment to social constructivism. The original expositions set out to investigate the hypothesis that *all* scientific knowledge is socially constructed to some extent, and possibly entirely so constructed. This claim certainly opposed most scholarship in history of science up to that time, but members of the Edinburgh School and others employed historical case studies as the basis of such research.

An early and highly influential example is the still controversial historical hypothesis put forward in 1971 by American historian of science Paul Forman. The 'Forman thesis' argues that the content of early quantum mechanics was shaped by the culture in which it developed, interwar Weimar Germany. The unexpected defeat of Germany in the First World War, it is claimed, caused a loss of confidence among the educated elite in rationality, deterministic processes and even causality itself. In this cultural environment, Weimar physicists opted to support the uncertainty principle proposed by Werner Heisenberg rather than alternate interpretations of quantum mechanics. As a result of these cultural pressures actively shaping the subject in the German context and its rapid international spread, the Copenhagen interpretation became the new orthodoxy. Forman's claims were followed by a generation of historical studies to further explore the social and cultural mechanisms that could influence the content of science.

Other variants of the sociology of scientific knowledge, notably attributed to French scholar Bruno Latour (1947–), argued from the 1970s for more radical understandings of science, technology, knowledge and technical products. His early work *Laboratory Life: the Social Construction of Scientific Facts* (1979, co-written with sociologist Stephen Woolgar (1950–)) unconventionally explored science by applying the methods of anthropology to a biology laboratory. These perspectives are beyond the scope of this book, but they, too, continue to motivate current research by some historians of science.

Such theorizing about the scientific enterprise may appear arid and divorced from the concerns of previous generations of historians of science, who more frequently adopted a narrative style (that is, constructing carefully researched stories of scientific episodes and historical figures). They may also fail to tempt the casual reader with their intellectual vistas. However, these radical positions, during the early 1990s, assumed a public and even political dimension. Dubbed the 'Science Wars' by the American media, the differences in viewpoint between radical constructivists and practising scientists were played out in magazine articles, campus debates and television interviews. In their crudest form, they illustrated a polar division between so-called 'relativists' and 'realists'. The relativists argue, with varying degrees of compromise, that scientific belief is influenced, shaped or determined by the society in which it is practised. The realists, calling upon older and still widely accepted philosophical foundations, argued that human knowledge based on rational scientific approaches is ultimately unlimited in its accuracy and power to describe the natural world. Both extremes accommodate nuanced approximations, making the 'wars' more a spectrum of discord. While the 'Science Wars' have cooled down, they are a potent illustration of the relevance of history of science in contemporary culture.

Anti-scientific movements and popular belief

The section above limited itself to changing scholarly opinion since the Second World War. But, as discussed in chapter 5, one of the most dramatic features of science over that period was the rise and fall of public confidence in scientific authority. So, alongside the scholarly turns, we can track changing understandings of science in popular culture.

Popular criticisms of science have roots as old as the scientific revolution, and have been supported by a range of scholars. During the second half of the eighteenth century, the grand aspirations of the Enlightenment were criticized by scholars such as Jean-Jacques Rousseau (1712–1778). Rousseau criticized the power and adequacy of rationalism to create a better world, and argued that humans inevitably were corrupted by society. He suggested that the advancement of knowledge had concentrated power in the hands of governments to the detriment of individual liberty.

Supported by such ideas, *Romanticism* became an important cultural and intellectual force influencing literature, art and music through the mid nineteenth century. The movement represented, in part, a resistance to Enlightenment claims. It stressed direct individual experience, imagination, emotion and intuition over cold rationality. While no consensus can be identified, Romanticism challenged the scope of reason and emphasized subjective human qualities. By extension, this challenged universal laws and scientific methodologies such as reductionism and quantification as a means of describing and explaining the complexities of the natural world. Incidentally, such perspectives inform some contemporary scientific concerns, too: environmentalism and so-called 'deep ecology' owe much to the Romantic movement, in opposition to technologically oriented solutions that can be linked more closely to the world views of the mechanical philosophers and many of the other practitioners that have been the focus of this book. In exchange for the expanding methods of science, proponents of Romanticism offered holistic, multi-layered description founded on particular experience.

Reductionism: the breaking down of a problem into more easily explainable parts, or the simplification and generalization of an explanation in terms of a more fundamental one. (Note: there is a distinct definition that may be encountered: in biology, where reductionism can refer to a materialist explanation of life.)

Holism: the consideration of multiple scales and interconnected contributions making up an effect.

The most influential scientific expression of this was *naturphilosophie*. Most widely supported in the German-speaking world by exponents such as Johann Wolfgang Goethe (1749–1832), this philosophy emphasized the interconnectedness of nature. Its approach promoted alternatives to the 'new science' of the seventeenth and eighteenth centuries, rejecting the procedure of dividing problems into more easily manageable portions. Goethe, for example, championed explanations of light and colour that were distinctly at odds with those of Newton a century earlier. A drawback of his colour theory was that, unlike Newton's, it was difficult to make predictions from it.

While the late nineteenth century witnessed a growing popular acceptance of progress as discussed in chapter 4, the pace of scientific and technological change provoked varied critical responses. Romanticism had offered an early alternative, but quite distinct counter-forces evolved in the twentieth century. Emerging in Switzerland during the First World War, *Dadaism* was a cultural movement that expressed a rejection of logic and reason. Through art, theatre, manifestos and design, Dadaists expressed irrationality and chaos as a reaction to the conformity and perverse 'logic' that they argued had led to war. Within a decade the movement had fostered *surrealism*. Surrealist artists, writers and performers juxtaposed unrelated and dream-like images, communicating a rejection of logic and orderly sequential thought.

In their very distinct ways, the Romantic, Dadaist and Surrealist movements were important examples of opposition to the techno-scientific basis of modern culture that was expanding during the nineteenth and early twentieth centuries. They challenged the completeness of scientific explanation and offered multiple perspectives in place of general explanation. Although Dadaism and surrealism were relatively elite and narrow in membership, they affected the wider public at least peripherally and temporarily.

A more direct expression of popular anti-scientific sentiments was through the adoption of non-Western religious, medical and metaphysical ideas. The New Age movement, for example, can be characterized as an individualistic approach to spiritual exploration and consciousness. Like the artistic movements described above, it criticizes the constraints and limitations of the scientific approach, and argues for a holistic understanding of the natural world. Although the term *New Age* circulated from the early 1970s, there are identifiable links with ideas that developed during the late nineteenth century, such as spiritualism and alternative medicines. Here, again, universal definitions cannot be constructed but the body of ideas draws upon a wide range of religious concepts from many cultures. Some concepts, such as meditation and reincarnation, have links to Eastern religions. An attention to mystical and mysterious dimensions of knowledge has roots in a number of world religions, including Christian and Jewish sects and Shamanism.

The collection of practices may also amalgamate a variety of medical traditions from other cultures. Some of these are ancient and widespread, such as acupuncture (China) and Ayurvedic medicine (India). Others represent new interpretations of old concepts, such as aromatherapy's adoption of ideas traceable to alchemy. Admittedly this sparse survey cannot do justice to the forms of knowledge that challenge science; this book highlights the twists and turns of the scientific perspective and can only sketch a background against which to contrast it.

There is one further aspect of these alternatives that can be mentioned, though. Some critical perspectives do not merely challenge conventional science: they sometimes have sought to incorporate and extend it. Some alternative medicines have recognizable scientific links, such as therapies based on magnets or light. New Age thinking is informed by certain sciences, notably aspects of psychology and ecology. Its interpretations of quantum mechanics, for example, draw connections between consciousness, causality and spirituality. New Age claims promote notions of knowledge (epistemologies) that extend beyond the methods and theories of science, but frequently make reference to them. Like spiritualism, one of its disparate roots, some New Age beliefs borrow from the terminology of science. Where spiritualists detected 'vibrations' from the spirit world, promoters of 'crystal therapies' may invoke 'resonances', 'energy levels' and 'recharging'; alternative therapists may refer to 'toxins'. Such adoption of scientific jargon with unconventional meanings has been described as pseudo-scientific. The criticism of practising scientists usually focuses on the lack of reliable evidence for such claims, and the lack of precision of their foundational ideas.

Pseudoscience: a body of knowledge that claims scientific authority without appropriate scientific methodology.

Not all of the counter-forces to modern science are to be found in older or non-Western traditions. Stalwart opposition to Darwinism, for example, has in recent decades been buttressed by (piecemeal) arguments drawn from the history and philosophy of science. Rather like supporters of phrenology in the early nineteenth century, certain supporters of creationism have attempted to construct a 'creation science' that adopts some features of scientific methodology. They may cite certain scientists as figures of authority for their claims (although most have credentials outside biology) or may point to inadequately explained observations as crucial refutations of evolutionary theory. The work of historians to reveal the complex history of many scientific claims may be used by creationists to hint that no scientific orthodoxy is safe and authoritative. Such pick-and-mix scholarship nevertheless is inevitably patchy and underlain by a clearly non-scientific foundation: that one particular theory – the Biblical account of creation – is beyond critical investigation and adjustment. By contrast, Darwinian evolution and Mendelian genetics have adapted to new empirical evidence.

The most serious challenge to the consensus of biologists was probably the set of claims made by Ukrainian agronomist Trofim Lysenko (1898–1976), who argued during the 1940s and 1950s for the inheritance of acquired characteristics (rather similar to the ideas of Lamarck some 150 years earlier). His methods of mutating crops by vernalization (acquisition of spring-hardiness of crops by exposing young plants to winter conditions) were carefully tested in other countries and found to be irreproducible. The rise of Lysenkoism, which promised higher crop yields, was supported in the Soviet Union by active suppression of Mendelian genetics until 1964. What had, for a time, been Soviet science was recast as pseudoscience.

Having distinct aims and memberships, other forms of opposition to specific scientific claims have become increasingly visible from the late twentieth century. The case of British opposition to the mumps-measles-rubella (MMR) inoculation during the late 1990s is a typical example. The case is interesting beyond the British context precisely because it became controversial only there; it begs the question of what conditions were remarkable at that time in that particular country. A small-scale study by a doctor suggested that the MMR injection could be correlated with subsequent emergence of autism in a small number of inoculated children. Unswayed by large-scale trials and statistical analysis – the methodology of modern medicine – many anxious parents rejected the perceived dangers of inoculation in the face of unverified anecdotal claims. This did not necessarily represent a blanket rejection of science and medicine, but often a construction of seemingly more acceptable explanations: for instance, that the vaccine would be safer if divided into three separate inoculations for mumps, measles and rubella, respectively; that the British Medical Association, National Health Service and Department of Health responded directly and unanimously to demands of their political masters, and so were suspect; or that 'rogue' doctors would routinely be attacked when they threatened industrial, institutional and professional establishments. Even for rational audiences, conspiracy theories may prove more compelling than plodding, and increasingly invisible and incomprehensible, science. Such claims are generally challenged by practising scientists who argue that, even taking political, social and cultural factors into account, bodies of expert and independent peers (operating in different countries under distinct political and religious systems, for example) tend, in the long run, to agree on questions of scientific fact.

These very recent challenges to scientific knowledge and practice are not unprecedented: as this book suggests, different approaches to knowledge have always coexisted. What makes them remarkable is their increased visibility in the past few generations after a long decline in the West since the scientific revolution. Old and new, they are topics of direct relevance to historians of science. Not only are they linked to ideas as old as modern science itself, but they can be expected to influence twenty-first century thinking about science and its integration into wider culture. The history of science, therefore, is firmly embedded in the analysis of the present-day.

Between the universal and the particular

We live in interesting times, to paraphrase an ancient Chinese blessing. History of science today is more vibrant and relevant than ever in the past. It is enriched – and made contentious – by other disciplines and perspectives. The debates of the 'science wars' have diminished considerably, but proponents of both extremes continue to lob volleys through historical studies. As a result, the hot spots in the history of science are equally changeable and multi-dimensioned. The field offers timely opportunities for research and exploration based on every type of human scholarship.

This *Beginner's Guide* has suggested that the observation, innovation in technique, logical reasoning and application of knowledge so characteristic of science are widely shared human attributes readily discernible across human societies. It has argued that natural phenomena are an inexhaustible resource to motivate human curiosity and research. Equally universal are the drives to explain patterns and to apply knowledge in the pursuit of power and control. The history of science focuses on the myriad contexts in which these human attributes have been expressed and shaped.

Alongside this seemingly ubiquitous drive, though, are dramatically different cultural expressions. The activities we call 'science' have emerged and mutated at particular times and places and been shaped and applied in those environments. The techniques of how best to recognize and weigh up these events and contexts have motivated – and sometimes divided – historians and scientists. The challenging and continuing goal for history of science is to detect and explain the subtle intellectual and cultural interactions of this diverse human enterprise.

Further reading

The books listed here are, for the most part, low cost editions and varied examples of recent work in the history of science, ranging from popular writing to scholarly but accessible monographs.

Chapter 1

Chalmers, A. F. 1978. *What is This Thing Called Science?* Milton Keynes, Open University Press.

Dobbs, B. J. T. and Jacob, M. C. 1994. *Newton and the Culture of Newtonianism*. Atlantic Highlands N. J., Humanities Press.

Kuhn, T, 1962. *The Structure of Scientific Revolutions*. Chicago, University of Chicago Press.

Olby, R. C., Cantor, G. N., Christie, J. R. R. and Hodge, M. J. S. (eds.) 1989. *Companion to the History of Modern Science*. London, Routledge.

Chapter 2

Crombie, A. C. 1995. *The History of Science from Augustine to Galileo*. New York, Dover.

Crowe, M. J. 2001. *Theories of the World from Antiquity to the Copernican Revolution*. New York, Dover.

Grant, E. 2001. *God and Reason in the Middle Ages*. Cambridge, Cambridge University Press.

Kelly, D. H. and Milone, E. F. 2005. *Exploring Ancient Skies: An Encyclopedic Survey of Archaeoastronomy*. New York, Springer.

Lloyd, G. E. R. 1999. *Magic, Reason and Experience: Studies in the Origins and Development of Greek Science*. Indianapolis, Hackett.

Moran, B. T. 2005. *Distilling Knowledge. Alchemy, Chemistry, and the Scientific Revolution*. Cambridge MA, Harvard University Press.

Sobel, D. 1998. *Galileo's Daughter: a Drama of Science, Faith and Love*. London, Fourth Estate.

Chapter 3

Browne, J. 2002. *Charles Darwin: The Power of Place*. New York, Knopf.

Darwin, C. 2006. *On the Origin of Species*. Mineola, Dover Thrift Edition.

Dear, P. R. 2001. *Revolutionizing the Sciences: European Knowledge and its Ambitions, 1500–1700*. Basingstoke, Palgrave.

Galilei, Galileo 2001. *Dialogue Concerning the Two Chief World Systems: Ptolemaic and Copernican*. New York, Modern Library.

Henry, J. 2002. *The Scientific Revolution and the Origins of Modern Science*. Basingstoke, Palgrave.

Kragh, H. 1996. *Cosmology and Controversy: the Historical Development of Two Theories of the Universe*. Princeton, Princeton University Press.

Miller, D. P. and Reill, P. H. 1996. *Visions of Empire: Voyages, Botany, and Representations of Nature*. Cambridge, Cambridge University Press.

Chapter 4

Dixon, T. 2008. *Science & Religion: A Very Short Introduction*. Oxford, Oxford University Press.

Gigerenzer, G. 1989. *The Empire of Chance: How Probability Changed Science and Everyday Life*. Cambridge, Cambridge University Press.

Gould, S. J. 1997. *The Mismeasure of Man*. London, Penguin Science.

Kevles, D. J. 1985. *In the Name of Eugenics: Genetics and the Uses of Human Heredity*. New York, Knopf.

Markham, A. 1994. *A Brief History of Pollution*. London, Earthscan.

Smith, C. 1998. *The Science of Energy: A Cultural History of Energy Physics in Victorian Britain*. Chicago, University of Chicago Press.

Stenhouse, J. and Numbers, R. L. 1999. *Disseminating Darwinism: the Role of Place, Race, Religion and Gender*. Cambridge, Cambridge University Press.

Chapter 5

Haber, L. F. 1985. *The Poisonous Cloud: Chemical Warfare in the First World War*. Oxford, Oxford University Press.

Hughes, J. 2003. *The Manhattan Project: Big Science and the Atom Bomb*. Cambridge, Icon.

Knight, D. 1986. *The Age of Science: the Scientific World-View in the Nineteenth Century*. Oxford, Basil Blackwell.

Krige, J. and Pestre, D. 1997. *Science in the 20th Century*. Amsterdam, Harwood Academic.

Reid, R. W. 1971. *Tongues of Conscience: War and the Scientist's Dilemma*. London, Panther.

Wang, J. 1999. *American Science in an Age of Anxiety: Scientists, Anticommunism and the Cold War*. Chapel Hill, University of North Carolina Press.

Chapter 6

Alic, M. 1986. *Hypatia's Heritage: A History of Women in Science from Antiquity to the Late Nineteenth Century*. London, Women's Press.

Iggers, G. G. 1998. *Historiography in the Twentieth Century: From Scientific Objectivity to the Post-Modern Challenge*. Hanover, University Press of New England.

Sayre, A. 1978. *Rosalind Franklin and DNA*. New York, Norton.

Schiebinger, L. 1989. *The Mind Has No Sex? Women in the Origins of Science*. Cambridge, MA, Harvard University Press.

Selin, H. 1997. *Encyclopaedia of the History of Science, Technology, and Medicine in Non-Western Cultures*. Dordrecht, Kluwer Academic.

Watson, J. D. 1999. *The Double Helix: A Personal Account of the Discovery of the Structure of DNA*. London, Penguin.

Chapter 7

Barnett, S. A. 2000. *Science, Myth or Magic? A Struggle for Existence*. St. Leonard's, Allen & Unwin.

Collins, H. M. and Pinch, T. J. 1998. *The Golem: What You Should Know About Science*. Cambridge, Cambridge University Press.

Hess, D. J. 1993. *Science in the New Age: the Paranormal, its Defenders and Debunkers, and American Culture*. Madison, University of Wisconsin Press.

Ladyman, J. 2002. *Understanding Philosophy of Science*. London, Routledge.

Latour, B. and Woolgar, S. 1979. *Laboratory Life: the Social Construction of Scientific Facts*. Princeton, Princeton University Press.

Ziman, J. M. 1998. *An Introduction to Science Studies: the Philosophical and Social Aspects of Science and Technology*. Cambridge, Cambridge University Press.

Index